COCIDO
MADRILEÑ
CON GARBANZOS
DE FUENTESAVCO

BEANS

Pulses,
A global journey
Has been prepared by the FAO Office for Corporate Communication

SEE
VIDEO
www.fao.org/
pulses-2016

Text: **Luis Cepeda** and **Saúl Cepeda Lezcano**.
Photographs: **Samuel Aranda** (Spain and Turkey), **Giuseppe Bizzarri** (Brazil), **Adam Wiseman** (Mexico), **Benjamin Rasmussen** (United States), **Alan Keohane** (Morocco), **Paul Joynson Hicks** (Tanzania), **Asif Hassan** (Pakistan), **Atul Loke** (India) and **Justin Chin** (China).

ISBN 978-92-5-109343-6

Pulses

A GLOBAL JOURNEY

Table of contents

Chefs

1
2
3
4
5
6
7
8
9
10

3

Nutritious seeds for a sustainable future

Our world today faces a tough challenge: ensuring food security while providing a balanced diet for everyone around the globe. The figures are daunting: around 800 million suffer from chronic hunger and roughly two billion live with one or more micronutrient deficiencies. At the same time, over half a billion people are clinically obese.

Overcoming hunger and malnutrition in the twenty-first century means increasing food quantity and quality, while making sure we produce food sustainably, efficiently and safely. In September 2015, world leaders adopted the 2030 Agenda for Sustainable Development, a plan of action for people, the planet and prosperity. It defines a list of Sustainable Development Goals aimed at ending hunger and malnutrition, as well as eradicating extreme poverty and tackling climate change, among other objectives.

The International Year of Pulses is helping to kick off the 2030 Agenda. Focusing on seeds for sustainability, FAO wishes to promote actions that will contribute to ending hunger while protecting the environment, the planet and its inhabitants.

Pulses have been an essential part of the human diet for centuries. Yet their nutritional value is not generally recognized and their consumption is frequently under-appreciated. Undeservedly so, as pulses play a crucial role in healthy diets, sustainable food production and, above all, in food security.

This book takes you on a voyage around the world to demonstrate how pulses are important historically and culturally, as reflected in today's cooking. We are honoured to present ten world-class chefs sharing their secrets on pulse dishes that are both traditional and tasty. We hope these recipes will entice you to try some or all of them and encourage you to include more pulses in your weekly diet.

Throughout the International Year of Pulses, we will continue our efforts in bringing together a wide spectrum of different actors, from diverse sectors, to enhance awareness about pulse production and consumption around the world. Our intention is to create synergies among our many and valued stakeholders and to lay the foundation for projects aimed at expanding the role of pulses in sustainable food production.

FAO is disseminating information on pulses through both print and multimedia, as well as hosting events at national, regional and global levels. Through these regional dialogues and global awareness campaigns, we aim to stimulate discussion and information exchange among civil society, farmers, the private sector, researchers, government representatives and policy-makers, among others.

Our team of nutrition experts is already compiling a food composition database of pulses as part of the FAO/INFOODS Analytical Food Composition Database, providing specific data on pulses, biodiversity and their relation to agriculture and processing.

This year marks the beginning of making pulses a household staple for those people and nations who may not know their incredible properties. At the same time, we have made good progress towards our goal of having pulses enriching the earth and sustaining entire populations. These superfoods have been nourishing people since long before historical records began.

So much still needs to be done to end world hunger and provide food security and nutrition for a swelling global population, expected to reach 9,7 billion by 2050. But one concrete, promising, sustainable and cost-effective opportunity lies within the smallest of seeds found in a multitude of plants: pulses – seeds for a sustainable future.

José Graziano da Silva
Director-General

Presence and power of pulses

Pulses are ancient, very ancient. They are a resilient grain seed that has existed for millions of years, a sort of wonder plant that grows in almost any condition and locality. It is thought that their domestication could pre-date maize. Found in all four corners of the earth, excluding only the poles and infertile deserts, pulses are grown even in regions with extreme hot and cold climes. For many, pulses provide the main source of plant protein. And they are not just good for you, they taste good too.

Across cultures and cuisines, their culinary versatility has given rise to a host of delicious recipes on every continent. Their main virtues are their vast geographical range, high nutritional value and low water requirements, their unique ability to self-fertilize (adding nitrogen to soils and improving crops along the way), along with maintaining their health benefits over a long shelf life. All these reasons make pulses an uncompromising enemy of hunger and malnutrition worldwide. Pulses are a genuine superfood for the future.

Air quality, access to clean drinking-water and a healthy diet are three big challenges for humanity's long-term survival. Dry pulses, combined with other staples, will be the key to meeting these challenges. As a category of foodstuff, pulses vary widely in nutritional properties and flavour, while as a whole, their unique qualities make them ideal for sustainable farming. And for good measure, from a cultural viewpoint, dry pulses are a symbol of travel, globalization and coming together. In ancient times, pulses kept the troops well fed, while in Italy, bags of pulses were served up to augur prosperity each new year. Domesticated pulse have been common to all peoples since olden times, with no distinction of race, religion or culture. If the need for food is something that unites all people, the extraordinary global reach of pulses is a powerful universal language between nations. Pulses originating in Asia are found in Africa and vice versa, while African varieties now grow in the Americas, and American strains have ound their way over to Europe and Oceania.

A brief guide to pulses

Pulses belong to the *Fabaceae* or *Leguminosae* family, these plants are the world's third-largest group of plant life. They are thought to have originated some 90 million years ago, with a diversification process beginning in the early Tertiary era.

People have cultivated pulses since farming began, and they were among the first plants in the world to be domesticated. Dried pulses are the dehydrated edible seeds of these leguminous plants that produce from one to twelve grains of various sizes, shapes and colours. Their seeds can be used for human consumption or animal fodder.

Pulses are eaten all over the world in stews, flours, purées, accompaniments, snacks and desserts. They are a rich source of protein and essential amino acids that act as the perfect complement to cereals. They are also a good source of carbohydrates and micronutrients, as well as high-quality dietary fibre. Their low fat content and the interaction of their sterols have been proved to be effective at maintaining low cholesterol levels and reducing blood pressure.

These hearty plants are not only healthy, they are also good for the earth. Leguminous plants have nitrogen-fixation properties, a natural attribute that is important for soil enrichment.

What does FAO consider as pulses?

Though taxonomically it is correct to include fresh peas, green beans, soybeans and alfalfa in this plant family, the FAO categorizes these as vegetables. Likewise, seeds that are grown for biofuel do not fall into the category of pulses as far as FAO is concerned.

Varieties of pulses

The following list illustrates the major groups

DRIED BEANS

Borlotti beans

Black beans

Adzuki beans

Cannellini beans

Red kidney beans

Haricot beans

Flageolet beans

Pinto beans

Mung beans

Urd beans (black gram)

Tepary beans

LUPINES

Lupines

BAMBARA BEANS

Bambara Beans

BROAD BEANS

Tepary Beans

Urd Beans (BLACK GRAM)

Adzuki Beans

Black Beans

Cannellini Beans

Flageolet Beans

Haricot BEANS (NAVY)

Borlotti Beans

Mung Beans

Red Kidney Beans

Pinto Beans

Broad Beans

8

LENTILS

Red lentils

Yellow lentils

Green
or brown lentils

Puy lentils

Umbrian lentils

Green and Brown Lentils

Yellow Lentils

Umbrian lentils

Red lentils

Puy Lentils

CHICKPEAS

Bambai chickpeas

Desi chickpeas

Kabuli chickpeas

GREEN CHICKPEAS

Kabuli

Bambai

DRIED PEAS

Dried green peas

Dried pigeon peas

Vetches

Dried cowpeas

Winged beans

Sword beans

Dried green peas

Dried Pigeon Peas

Vetch

Dried Cow Peas

Winged Beans

Sword Beans

9

Caring for & cooking your pulses

Keep in mind

—● When legumes are not pre-soaked or they are more than 18 months old, two parts of water in weight or volume should be used to one part of pulses (also called legumes). If they were soaked in advance, one somewhat generous portion of water is sufficient to compensate for evaporation when they are cooked over medium heat. In any event, when cooking in an open pot you must always make sure that the liquid completely covers the legumes, and add water if needed; hot water if cooking chickpeas, and cold or lukewarm water for other pulses.

—● If the legumes were not pre-soaked, they can still be softened by covering them with cold water and adding one-half teaspoon yeast for each half kilogram of legumes. After cooking over low heat for 40 minutes, drain and add cold water (less water if cooking chickpeas) to start the stewing process.

—● Chickpeas are the only legume that should be cooked – after pre-soaking them for no less than eight hours – by placing them in hot or boiling water instead of placing them in cold water, which is what is done with other legumes.

—● If they are cooked in a pressure cooker, this precaution should also be taken and they should be placed into the pot with hot water. Furthermore, to prevent them from breaking or losing their skins when they bump into each other during boiling, they should be placed in a strainer before putting them into the pot.

Storage

—● Most dry legumes can be stored for a long time, even for years, without spoiling and still retain their nutrients. Pulses will retain excellent quality for 18 months, although the longer they are stored, the more time-consuming and expensive it is to cook them. Cooks generally prefer fresh legumes when they are in season, which in the northern hemisphere begins in September, and in the southern hemisphere, around March.

—● They should be stored dry, preferably in airtight glass containers. When selecting, be sure to read labels and choose those that do not have added flavors and those with the lowest sodium content possible, in order to be able to season them to your own taste. In some cases, taste may be modified by the additives used during the preservation process. Storage should be in a cool, dry place. If pulses are exposed to a somewhat warm environment, there is the same danger of food poisoning, as with any other seasoned food.

Condiments

—● One bay leaf is often added to cooked lentils, which gives them a delicious aroma and flavour. Another classic among seasonings is a mix of bouquet of aromatics, called a bouquet garni in professional cooking, comprised of thyme, parsley and bay leaf, ideal for stews and casseroles. Special seasonings such as whole pepper corns, cloves or curry, provide unusual savory flashes of taste, boldly fusing with vegetables, such as okra in Cajun cooking. Masalas or tahini offer yet another taste sensation to your pulses.

Soaking

The first step in prepping legumes [be]fore cooking is to wash them, as they [migh]t contain impurities that should be [elim]inated, such as the remains of the [stem] or the chaff, dirt, small stones, or [smal]l seeds that will not soften during [cook]ing. This is a precaution that is [increa]singly less necessary, because [pac]kaged legumes go through an [effec]tive pre-selection process. Soaking [shou]ld ideally not last more than 12 [hour]s. After this time, the water should be [chan]ged. Soaking legumes for 24 hours [can] start the fermentation or sprouting [proc]esses. After soaking, rinse them [until] the water runs clear to eliminate the [sug]ars that come loose during soaking, [mak]ing pulses difficult to digest.

Cooking

In general, legumes are quite easy to cook.
1) Place them in a pot covering them with plenty of water and ingredients that add flavor.
2) Heat the water till boiling, and then
3) Cook on low heat until soft.

To stew legumes, place (prepared) pulses into your pot and bring to boil as above. They should be cooked over low heat with a little olive oil and a vegetable such as onion, garlic, leeks or shallots, as seasoning.

For bean or lentil soup, place them in a pot and cover with cold water or broth; do not salt in order to keep them from hardening. To make them more tender, once the legumes

have just started to boil, you can change the water for cold water, or you can shake the pot up to three times: adding a splash of cold water to decrease the boiling. When removing legumes, especially beans, swirl the contents of the pot by picking it up by its handles rather than using a spoon, which will prevent the legumes from breaking. In truth, this only affects the aesthetics of the dish, but to many cooks, that is equally important! Salt to taste and serve.

Chickpeas are an exception to the way legumes are usually cooked. As previously stated – but it bears repeating – start with warm water and a little salt, and if more water needs to be added, always add hot water. Adding in cold water – which is fine for other legumes – causes chickpeas to undergo a brusque temperature change that interrupts the cooking, causing them to harden thus preventing them from cooking uniformly.

Cooked legumes should not be left at room temperature for more than four hours. It is best to keep them above 55°C so that they do not develop any type of bacteria, and they should be refrigerated or frozen as soon as possible after cooling. If they are going to be used in salads, add vinegar or lemon zest to keep bacteria from growing.

Utensils

Pots, pans or casserole dishes should be made of stainless steel, or they should be enamel-coated. Wooden spoons or utensils should be used for stirring, because handling or simple contact with metal instruments during cooking can deteriorate, break or remove the skin off the legumes.

Wide casserole dishes can be used in which the legumes just cover the bottom of the pan. This will allow them to cook uniformly, with the added advantage that the legumes resting on the bottom do not bear the weight of those on top.

A pressure cooker is very efficient, especially for shortening cooking times, although some expert chefs are hesitant to use them. Except for the risk of opening them too soon when there is still pressure inside – nearly impossible today with more modern pressure cookers – this type of cooking does not spoil legumes in any way. In fact, pressure cookers perfectly concentrate all the pulse qualities, more so than cooking them in an open pot. Pay close attention to each manufacturer's instructions on cooking times and temperatures so that your legumes don't turn into mush.

Nine Benefits of Pulses

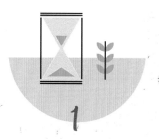

1

2

3

4

Boast a long shelf life

When stored in airtight containers, pulses can last months, even years, without spoiling. For subsistence farmers, this could mean the difference between life or death should they suffer a bad harvest or natural disaster (like floods) that wipes out their entire harvest.

Keep you healthy

FAO recommends that people eat at least 400 g of fruit and vegetables per day, which includes pulses and other legumes. This is equivalent to eating about 25 g of dietary fibre per day. As pulses are high in dietary fibre, they can help prevent obesity, reduce blood pressure and reduce the risk of heart disease.

Good news for poor farmers

Growing pulses can mean a variety of benefits for poor farmers. Pulses grown with other crops or in rotation will fertilize the soil, and can increase yields on less productive farmlands. Some pulses, such as beans, fetch more at the market than some cereals giving poor farmers a better chance at ending the poverty cycle.

Help other crops to grow

Pulses' nitrogen fixing qualities mean that crops planted alongside pulses reap the benefits and grow faster. Pulses are also deep rooting, which means they do not compete with other crops for water. This makes them ideal companions.

Cost less to grow

Plants need nitrogen fertilizers to grow. Pulses can fix their own nitrogen in the soil, which means they nourish the soil instead of depleting it. This means farmers don't have to buy nitrogen fertilizers, which in poor areas is a substantial cost savings for the farmer.

'Clean' crops: do not emit greenhouse gases

In stark contrast to animal products, pulses have been shown to emit hardly any greenhouse gases (lentils emit 0.9 percent). Cultivating and eating more pulses would bring huge benefits to the environment.

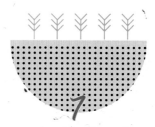

Help fertilize soil

While other crops deplete the terrain in which they grow, pulses actually do the opposite. Grown as green manure or cover crops or as forage for livestock, pulses can build up nitrogen in the soil even faster; fixing as much as 300 kg of nitrogen per hectare. Pulses also release hydrogen gas into the soil – up to 5 000 litres per hectare per day, exerting yet another positive impact on soil biology.

Zero waste

Every part of the pulse can be used. The pods can feed people, the shoots used for animal feed, or the pulse can be left in the earth to provide nourishment for the soil.

Need less water to grow

Seventy percent of the world's accessible fresh water is used for agriculture. Twenty-seven percent of the world's water footprint comes from the consumption of animal products. Pulses need 20 times less water than animal products to grow. In industrialised countries, moving towards a vegetarian diet can reduce our food-related water footprint by a full 36 percent.

A world of pulses

North America

THE SUPERFOOD OF THE FUTURE

Although North America is home to two developed countries where meat and fish protein are preferred by consumers, these are also countries with exceptional farming models. The culture of pulses in Canada and the United States is closely linked to their extraordinary production processes and export trade. Canada alone exports lentils, beans and chickpeas to 150 markets worldwide, with the main producing regions being Alberta, Manitoba, Quebec and, above all, Saskatchewan. Much of this is largely due to the important role of tinned pulses, especially baked beans. Baked beans are a culinary feat not only in terms of how they are preserved, but also for how well they can travel. They have even achieved cultural icon status through cinema, television, literature and in many art forms. Just think of Andy Warhol's *Black Bean*.

We should also keep in mind that these are two countries with powerful multicultural societies. And this cultural plurality is naturally reflected in the food found in North American cities. Canada, for example, has seen significant immigration from countries that consume a high proportion of pulses, such as India, the birthplace of 12 percent of foreign residents in Canada. Meanwhile, cities like Toronto are home to 90 different nationalities. It is not surprising, then, that the world's second largest country provides a warm hearth for food from Brazil, France, India, Iran, Italy, Mexico, Portugal, Spain and Turkey and all the flavours they bring with them from spicy tarka *daal*, to hearty *feijoada*... To this mix we can add Cajun specialities from Acadia (the old colony of New France), also found in the southern United States. Their unique cuisine is based on pulses, rice, seafood and the "holy trinity of vegetables", namely celery, onion and peppers.

For its part, the United States offers an immense cultural variety of foods across its territory, due to the complex history of its formation and subsequent waves of immigrants throughout each state and region. Although it comes as no surprise that major cities like Chicago, Los Angeles and New York are home to a wide range of ethnic cuisines, it is important to note that dried pulses consumption varies widely from place to place. Migration has clearly played an important role, which explains why bean soups in Idaho have their origins with Basque immigrants. Influence from the Canary Islands shows up in pulse stews in Louisiana, a state known for its veritable creole gastronomy with New Orleans at its heart and strongly influenced by France, Spain and Latin America.

The south's *soul food* is equally important in the gastronomy firmament, using a host of dried peas. Bean soups from Appalachia and traditional Midwest and northeastern dishes feature dried pulses (once vital to old mining communities for their shelf life) and have left their mark in recipes today. And of course, honourable mention must go to Native Americans and their time-old tradition of beans. They were the original inhabitants of the Eastern Woodlands, the Great Plains and the regions of the Gulf of Mexico. With them, southerners who were most influenced by Central America and the Caribbean. Cornbread is common to both as well as succotash, made from a mixture of maize and beans.

Of course, the influx and influence of Latinos in the United States cannot be ignored – and not because of the remarkable success of restaurant chains based on Latin American cuisine. Around 13 percent of the US population is Latino, spread across the entire country, from California to Florida and from Houston to New York. For the most part, they are of Mexican origin, but there are also Cubans, Puerto Ricans and, to a lesser extent, groups from Central and South America. But common to all of their gastronomic traditions are dried pulses, especially beans (known as *frijoles*). It goes without saying that the progressive growth in the Latino population has brought about a gradual change in domestic eating habits in North America. Thanks also to recent recommendations regarding a healthy diet, pulses will continue to take centre stage and animal proteins will be replaced by foods rich in vegetable proteins.

United States of America

18

RON PICKARSKI'S ECO-CUISINE

Ronald A. Pickarski has become a rising star for vegetarian chefs specialising in pulses. He combines pulses with spices and wheat flour, creating a vegetarian meat substitute, turning this into prime veggie burgers, the taste and texture of which resemble meat in the mouth. He works his magic turning black beans, quinoa and sun-dried tomatoes into a delicious and healthy bread. Pickarski is deeply interested in the proteins present in pulses, since these are the most important molecules in terms of an organism's structure. They even regulate body mass, which is why he wants to make sure that pulses are an integral part of our diet in the future. He uses statistics sensitively and consistently, citing the figure that by 2054 the global population will need approximately one billion tonnes of protein, and alternative sources such as pulses can provide one third of this amount.

But Pickarski was not always in the culinary limelight. He cut his teeth (literally) between 1968 and 1993, as a monk in the Franciscan Order of Friars Minor in Oak Brook (Illinois). He worked in the kitchens and began to develop specially-designed diets using the organic food grown on site, before moving on to preach cookery and become a nutritional consultant. His religious affiliation is rather unusual in chefs, but it is a source of great pride for Pickarski, and in all likelihood is at the root of the humility and honesty he shows in all of his work with food.

Currently living in Boulder (Colorado), Ron is the Executive Chef of Eco-Cuisine Inc., the company he founded in 1993. They provide consultancy services to promote healthy eating, research natural foods, develop products based on plant proteins and organize gourmet vegetarian cooking classes. He holds Certified Executive Chef status from the American Culinary Federation, and in 1994 he also founded American Natural Foods in Boston (Massachusetts), a non-profit organisation with the mission to inform businesses and the public about vegetarianism and promote plant-based products, among which pulses are key. "Pulses," says Pickarski, "are an ancient foodstuff and perfect for human consumption. They are an important source of protein, they are bursting with complex carbohydrates and fibre, and what's more they're low in fat. Only soya has a high fat content. I believe that pulses should become a main course for everybody, and my philosophy regarding pulses is that they should be seen more and more as a substitute for meat."

He has dedicated years of his life to developing a cuisine based on plant products, collecting tips for healthy eating and taking part in the International Culinary Olympics. His mission is to raise vegetarian cooking in general, and very specifically cooking with pulses, to the level of classic gourmet cuisine. It is clear to Pickarski that although the United States both produces and consumes a lot of pulses, they are still not generally

well-perceived in culinary circles. Although most people would agree that a vegetarian diet is healthy, they do not consider pulses to be more important than fresh vegetables. Nevertheless, there are dishes that have helped pulses to be accepted by the more reluctant: trailblazers such as Boston baked beans, Cajun *gumbo* with *alubia criollo* beans in Louisiana or pinto bean *burritos* in the southeast. Appearing more often on menus, they suggest that not only are there traditional recipes with pulses in North America, but new and exciting dishes, too.

Pickarski's incredible output has included designing, opening and promoting many innovative restaurants in Florida, Massachusetts, Illinois, Kansas, Wisconsin and Colorado, showcasing vegetarian and vegan food, a macrobiotic diet and even

pulses themselves. He has also sung the praises of pulses in books, videos, and on TV, featuring them on *Home on ABC*.

Recently, Pickarski published *The Classical Vegetarian Cookbook*, a modern compendium of responsible and ethical food. Alfonso Contrisciani, perhaps the North American chef who most actively supports sustainability, calls it a "Valuable and necessary tool for all who want to fully understand this new 21st century cuisine – animal-free with classical standards and classical taste." In its some 400 recipes, the book explores vegetarianism, making a sound case for a vegetarian diet, tackling head-on the problem of protein sources. In it, one is inspired to create pulse dishes to satisfy modern tastes while guaranteeing a more sustainable future for our food.

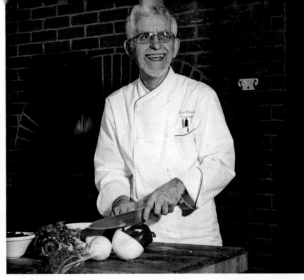

The chef Ron Pickarski in the kitchen of the University of Colorado, Boulder, United States of America.

Cannellini Bean Polenta Loaf

CANNELLINI BEAN POLENTA MAKES A COMPLETE PROTEIN WITH THE BEAN-GRAIN COMBINATION. IT IS AN ENTRÉE THAT NEEDS ONLY A SAUCE AND VEGETABLE TO SERVE AS A WHOLE MEAL.

3/4 CUP (170G) FINELY DICED ONIONS

3/4 CUP (170G) FINELY DICED RED BELL PEPPER

1 TABLESPOON MINCED FRESH GARLIC

2 TABLESPOONS EXTRA VIRGIN OLIVE OIL

1. PLACE THE OIL IN A 3-QUART SAUCEPAN; ADD THE ONIONS, RED BELL PEPPERS, GARLIC, CILANTRO, FENNEL, AND SALT.

2. SAUTÉ OVER MEDIUM HEAT FOR 8 MINUTES, OR UNTIL ONIONS ARE TRANSPARENT. ADD THE WATER AND CORNMEAL. BRING TO A LOW SIMMER AND COOK FOR 15–20 MINUTES, OR UNTIL MIXTURE IS SOFT AND THICK.

3. STIR IN THE BEANS AND OLIVES.
TRANSFER MIXTURE TO A GREASED 2-QUART LOAF PAN;
COVER AND LET SET FOR 30 MINUTES.

4. TO SERVE, CUT IN 1/2-INCH OR THICKER SLICES AND
SERVE WITH A MIX OF VEGETABLES FOR A COMPLETE
MEAL.

*CORN GRITS CAN BE SUBSTITUTED FOR CORNMEAL.
BECAUSE CORN GRITS MUST COOK ABOUT 30 TO 40
MINUTES AND THE WATER RATIO AND COOKING TIME
ARE DOUBLE THAN THAT OF CORNMEAL, DIRECTIONS FOR THE
RECIPE WILL CHANGE. IF SUBSTITUTING CORN GRITS
FOR POLENTA, USE AN EXTRA CUP OF WATER AND COOK
AN ADDITIONAL 10 TO 20 MINUTES UNTIL POLENTA IS
VISCOUS OR OF THE CONSISTENCY OF TRADITIONAL
CORNMEAL POLENTA.

1/2 CUP (115G) PITTED AND CHOPPED
KALAMATA OLIVES

1 TABLESPOON GROUND FENNEL SEED

2 CUPS (450G) CANNELLINI BEANS
(OR DARK RED KIDNEY BEANS FOR COLOUR)

1-1/2 CUPS YELLOW CORNMEAL
OR CORN GRITS *

3-1/2
CUPS WATER

1 TEASPOON
SALT

1/4 CUP (55G)
CHOPPED
FRESH CILANTRO

MEDITERRANEAN HUMMA-NUSHA

SERVES 3-4

35 MINUTES IF BEANS ARE PRE-COOKED.

This unique dish is a blend between a hummus and baba ganoush. It is fusion Mediterranean, offering a refreshing new flavour combination with a textured mouthfeel. Serve as an appetizer or in a sandwich.

1/2 CUP (115G) CHOPPED ROASTED OR GRILLED EGGPLANT

1/2 CUP (115G) CHOPPED ROASTED RED BELL PEPPER

1 TABLESPOON TAMARI (GLUTEN-FREE)

2 CUPS (450G) COOKED GARBANZO BEANS

1/2 CUP (115G) CHOPPED FRESH CILANTRO

TAHINI 1/4 CUP (55G)

LEMON JUICE

1 TEASPOON MINCED GARLIC

1. Place all ingredients in a blender or food processor and process until mixture is nearly smooth.

2. Serve as a dip or use 1/4 cup per sandwich.

CLASSIC CUBAN-STYLE
Picadillo Sauce
WITH BLACK BEANS

SERVES 5 — 20 MINUTES PREP AND 30 MINUTES COOK TIME WITH PRE-COOKED BEANS

SAZÓN
(THE DRY MIX)

1 1/2 TEASPOONS OF THIS MIX EQUALS ONE PACKET OF COMMERCIAL **SAZON**.

1 TABLESPOON GROUND CORIANDER
1 TABLESPOON GROUND CUMIN
1 TABLESPOON GROUND ANNATTO SEEDS OR PAPRIKA
1 TABLESPOON GRANULATED GARLIC
1 TABLESPOON SALT

I CHOSE THE PICADILLO SAUCE WITH BLACK BEANS TODAY, MARCH 21, 2016 IN HONOR OF OUR PRESIDENT MAKING THE HISTORICAL FIRST TRIP TO CUBA SINCE BEFORE THE CUBAN REVOLUTION. FOOD IS A UNIVERSAL FORM OF CELEBRATION AND PULSES ARE A UNIVERSAL INGREDIENT. TODAY IS A GOOD DAY FOR CUBAN-AMERICAN RELATIONS, THE UNITED NATIONS, AND THIS RECIPE COMMEMORATES THIS HISTORICAL DAY.

1. HEAT OLIVE OIL IN A SKILLET OVER MEDIUM HEAT; STIR GARLIC, ONION, AND GREEN BELL PEPPER INTO THE HOT OIL AND COOK UNTIL SOFTENED, 5 TO 7 MINUTES.
2. ADD PRE-COOKED BEANS AND COOK ON MEDIUM HEAT FOR 4 TO 5 MINUTES.
3. ADD OLIVES, RAISINS, CAPERS, TOMATO SAUCE, SEASONING, CUMIN, SUGAR, AND SALT TO THE MIXTURE.
4. COVER THE SKILLET, REDUCE HEAT TO LOW, AND COOK UNTIL THE MIXTURE IS FULLY HEATED, ABOUT 10 MINUTES.

1 TABLESPOON **SEASONING** →

1/2 CUP (115G) RAISINS

1/4 CUP (55G) QUARTERED AND PITTED GREEN OLIVES

2 TEASPOONS GROUND CUMIN

1/2 CUP (115G) CHOPPED GREEN BELL PEPPER

1 TABLESPOON OLIVE OIL

1 CUP TOMATO SAUCE

2 TEASPOONS MINCED GARLIC OR MORE TO TASTE

2-1/2 CUPS (560G) COOKED BLACK BEANS

1 TABLESPOON CAPERS

3/4 CUP (170G) CHOPPED ONION

1 TEASPOON WHITE SUGAR

SALT TO TASTE

*PICADILLO (PEE-KAH-DEE-YOH) IS A SPANISH WORD TO DESCRIBE A CENTRAL AMERICAN AND CARIBBEAN DISH OF GROUND PORK AND BEEF OR VEAL, ONIONS, GARLIC, AND TOMATOES USED AS A STUFFING (IN MEXICO) OR SAUCE (FOR BEANS IN CUBA). IN CUBA IT IS A CLASSIC RECIPE FOR BLACK BEANS OR GROUND BEEF AND IS TYPICALLY EATEN OVER WHITE RICE OR USED AS A FILLING FOR TACOS OR EMPANADAS. IT'S DELICIOUS WITH FRIED RIPE PLANTAINS. YOU CAN USE THIS SAUCE WITH ANY BEAN. OTHER APPLICATIONS INCLUDE CUBAN GRILLED TEMPEH OR TOFU PICADILLO.

Central America and the Caribbean

Mayan legend has it that a poor farmer was approached by Kisin, an evil being in Lacandon mythology, who told him that in seven days his soul would be taken to hell, but that for each of his remaining days he would be granted a wish. The ingenious farmer asked him, in the following order, for money, health, power, food and to travel and fulfil his dreams. On the seventh day, he asked Kisin to help him wash the black beans until they were white. This was an impossible task, since only one variety existed at the time. It is said that Kisin, dismayed, created beans in every colour so that he would never be tricked again.

Allegories aside, recent international studies like those carried out at the Università Politecnica delle Marche in Italy, place the origin of the bean, the *Phaseolus vulgaris*, in Mesoamerica and not in the Andean region as previously thought. This would help to explain the undeniable influence of beans in pre-Columbian cultures such as the Olmec, Maya, Aztec, Mexica, Mixtec, Tarascan, Teotihuacan and Zapotec civilisations, to name the most significant ones. Radiocarbon tests have made it possible to determine that certain spontaneously-growing beans found at archaeological sites are around 10 000 years old. It is known that in the seventh century BC, crops were already grown in the region, making them older than maize and one of the most ancient foodstuffs documented in the history of humanity. The wide variety of beans – whether in colour, size or shape – reveals the biological diversity that must have already existed at the time. It also lends weight to theories that these crops would have been traded between cultures, and even used as currency or for taxation. The records of the *Codex Mendoza*, for instance, show levies received by the Triple Alliance of the Valley of Mexico in the form of beans.

The importance of beans in the indigenous diet, along with the diversity of varieties, were chronicled by religious and military officials involved in colonising the region. Captain Gonzalo Fernández de Oviedo Valdés, for instance, in his *Historia general y natural de las Indias, islas y tierra firme del mar océano* (General and natural history of the Indies, islands and mainland of the ocean), a sixteenth-century work, describes the rich plant life of Central America. It highlights the importance of pulses, and in particular, their significance on the isthmus and in the Caribbean (where Christopher Columbus himself found strange crops, "very different beans to our own"). Needless to say, pulses continue to be a central part of the culinary traditions of the Antillies, Cuba, the Dominican Republic, El Salvador, Honduras and Nicaragua. It is also accepted among food historians that following the arrival of the Spanish in the Americas, beans began to cross the ocean. Spain proved particularly receptive in terms of cultivation and culture, and soon they spread throughout Europe. In return, Europeans brought chickpeas to the Americas, and although they never became as prevalent as beans, they adapted very well to the climate of northern Mexico; the region's leading producer and exporter of this pulse (known as *chícharo*) today.

Mexico, with a population of well over 100 million, was also the first country to be granted UNESCO Intangible Cultural Heritage status for its cuisine, mainly due to certain culinary methods that have survived since ancient times, and ingredients such as maize, chilli and beans. This pulse is known in the ancestral language of *Na'huatl* by the generic name *etl*, a word for beans found in many culinary references handed down through the ages.

The ancient bond between Mexicans and beans has resulted in their being featured in over half of the country's national dishes, extending across each and every region. Beans are such an important part of the Mexican diet that, despite the huge volume of production, the bulk of the crops grown are for domestic consumption. In contrast, lentils, another pulse prevalent in traditional Mexican cuisine, are primarily grown in the states of Michoacán and Guanajuato, but the local population only consumes a small proportion of national demand.

MEXICO CITY (MEXICO)

Chef Ricardo Muñoz Zurita in his restaurant *Azul Condesa*.

Mexico

RICARDO MUÑOZ ZURITA, BUILDING AN ENCYCLOPAEDIA OF BEANS

Since the Aztec Empire, Mexico has had a culinary history continuing up to today. The country's farming traditions and unusual growing techniques, along with its sensitivity to infuse ancestral influences, are all factors that have carried Mexican cuisine to be recognized as a World Heritage in 2010.

The declaration came about thanks to an initiative promoted by Mexico's Conservatory of Gastronomic Culture, a private institution that first envisaged the possibility of a national cuisine obtaining World Heritage status. Years earlier, five Mexican chefs led by Ricardo Muñoz Zurita paid a visit to UNESCO headquarters in Paris. To make their case, they brought with them a virtual farmer's market – 40 kilos of authentic Mexican produce: *escamoles, huitlacoche* and *chapulines; Puebla,* black and yellow mole; 120 kinds of chilli, *epazote, achiote* and *quelites;* pumpkins and squash, black *sapote* pulp, and no fewer than 50 native varieties of beans. "Taking our basic staples of maize, chilli and beans, combined with pre-Hispanic cooking methods, were the reasons for insisting," says chef Muñoz Zurita. He did not back down when faced with committee scepticism over what to do with a 40 kilo dossier.

For Muñoz Zurita, Mexico's and arguably Latin America's undisputed top chef, beans are synonymous with Mexican cuisine. He is well aware that times change, and with social transformation and global trends come new urban eating habits. But while rural traditions are often relegated to the dust bin of the past, in Mexico, beans remain a daily staple in a country where over 100 varieties are produced.

Muñoz Zurita points out that the *milpa* – the ancestral agro system of Mesoamerica that produces bean and maize crops – is consistent with systems developed by other civilisations around the world. In other words, soya and rice in the East, beans, chickpeas and wheat in Africa, beans and maize or wheat in the Americas, and beans, chickpeas, lentils and wheat throughout Europe owe their growth to early Mexican farmers.

29

Mexico

It is widely known that natural growing cycles and eating habits have led us to eat pulses and cereals together, thereby increasing the nutritional potential of proteins. This practice produces a synergistic effect that, in the case of beans, triggers an abundance of lysine, an essential amino acid that, as the chef states, "Aids collagen formation and calcium absorption in the body, maintains the nitrogen balance in adults, is very useful in the production of antibodies and stimulates the growth hormone."

Ricardo Muñoz Zurita was born in Mexico City in 1966. He was named the Prophet and Preserver of Culinary Tradition by *Time* magazine in 2001, one of many distinctions, like his membership in the Académie Culinaire de France. After gaining his first experience in his parents' family restaurant, he trained as a chef at San Diego Community College in California, Le Cordon Bleu in Paris, and the Culinary Institute of America in New York. He is the founder and manager of *Azul y Oro Café* in Mexico City at the university's Cultural Centre, and owns two leading centres of Mexican cuisine, the *Azul Condesa* and *Azul Histórico* restaurants, housed in unique premises in the capital's historic centre. He has authored a vast number of culinary publications, including the monumental *Diccionario Enciclopédico de Cocina Mexicana* (Encyclopaedic Dictionary of Mexican Cuisine), published in 2013; the product of 22 years of work. It devotes over 100 entries to beans alone. Even so, he says, this "falls short" of reality, given the endless possibilities of beans in Mexican cuisine.

Muñoz Zurita insists that more pulses should be included in our diets, particularly in large cities, where new, global, standardised and foreign habits have taken root. The nutritional properties of beans are unquestionable, while access is readily at hand: "It's quite easy and economical to buy dried pulses," he insists. He does admit, however, that a busy modern lifestyle isn't suited to the slow-cooked stews in which pulses often feature. The absence of a family member whose role is to be in the kitchen means that, for many, beans are starting to be seen as a nostalgic reminder of homemade cuisine. To counter this, Chef Muñoz Zurita praises the pressure cooker and timers on cooking equipment, which have sped-up the process and made timing more precise. "Another useful alternative," the chef suggests, "is pre-cooked or powdered pulses, prepared by reputable brand names and preserved perfectly in cans, sealed, or vacuum-packed – retaining the pulses' nutritional properties – while making our lives easier and making room for pulses."

Chef Muñoz Zurita buying ingredients at the local market of Medellín and in his restaurant, *Azul Condesa*, preparing a bean salad.

Frijol Colado

Place water, beans, onion and garlic in a pressure cooker on high heat for 1 hour.

When the safety valve begins to hiss, remove from heat and leave to cool until the safety valve has gone down.

Remove lid and check that the beans are tender.

Return the pan to medium heat, add salt, and cook for 10 minutes uncovered, then remove from the heat.

Drain beans, making sure to keep the cooking liquid aside separately.

Discard onion and garlic and place half the bean liquid and beans in a blender.

Blend to a thin, smooth consistency that passes through a strainer leaving almost no pulp, and set aside.

INGREDIENTS

2 LITRES WATER

400 G BLACK BEANS, SOAKED 12 TO 24 HOURS IN 4 CUPS OF WATER AND THEN DRAINED

1/4 MEDIUM WHOLE WHITE ONION (60 G)

3 LARGE GARLIC CLOVES, PEELED AND HALVED (12 G)

1 LEVEL TABLESPOON SALT (20 G)

FOR THE FRIJOLES COLADOS

2 TABLESPOONS VEGETABLE OIL (30 ML)
1/4 WHITE ONION, FINELY CHOPPED (60 G)
6 EPAZOTE LEAVES, FINELY CHOPPED
1 ROASTED WHOLE XKATIK CHILLI* (50 G)
BLENDED BEAN MIXTURE

HEAT OIL IN A SAUCEPAN ON HIGH HEAT UNTIL IT STARTS TO SMOKE.

ADD ONION AND FRY UNTIL TRANSLUCENT.

ADD THE EPAZOTE, XKATIK CHILLI AND BLENDED BEANS AND MIX, COOKING OVER LOW HEAT FOR 20 MINUTES UNTIL THE BEANS THICKEN, WITHOUT LETTING THEM DRY OUT.

SALT TO TASTE, REMOVE FROM HEAT AND SET ASIDE.

THIS PREPARATION IS A BASE FOR MANY RECIPES, SUCH AS PANUCHOS AND PAPA NEGRO, AND CAN ALSO BE SERVED WITH MEAT, FISH OR AS A MAIN COURSE. IT CAN BE PREPARED WITH CANNED BLACK BEANS TO SHORTEN PREPARATION TIME.

* YELLOW AND ELONGATED, ALSO CALLED GÜERO CHILLI

REFRIED BEANS

serves 6

1 CUP COOKED BLACK BEANS, BLENDED WITHOUT LIQUID (240 G)

+ FRIJOL COLADO
SEE RECIPE

PREPARATION:

1. HEAT THE OIL IN A FRYING PAN OVER A HIGH HEAT UNTIL IT BEGINS TO SMOKE

2. ADD THE FRIJOL COLADO CAREFULLY SO THE OIL DOES NOT SPIT, AND FRY FOR 3 MINUTES STIRRING CONTINUOUSLY WITH A WOODEN SPOON TO PREVENT THE FRIJOL FROM STICKING.

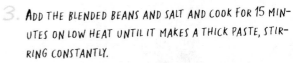

1/4 CUP CORN OIL (60 ML)

1 LEVEL TEASPOON SALT (7 G)

3. ADD THE BLENDED BEANS AND SALT AND COOK FOR 15 MINUTES ON LOW HEAT UNTIL IT MAKES A THICK PASTE, STIRRING CONSTANTLY.
SALT TO TASTE, REMOVE FROM HEAT AND SERVE HOT.

4. REFRIED BEANS ARE BEST SERVED NICE AND HOT ON A LARGE PLATE ACCOMPANIED WITH QUESO FRESCO, XNI-PEC SAUCE AND FRESHLY MADE CORN TORTILLAS.

Takgswayajun

2 CUPS
BLACK BEANS,
SOAKED OVERNIGHT
(412 G)

PLACE WATER, BLACK BEANS, ONION, GARLIC AND SALT
IN A PRESSURE COOKER ON HIGH HEAT.

AS SOON AS THE SAFETY VALVE BEGINS TO HISS, LOWER THE HEAT TO A MINIMUM
AND CONTINUE TO COOK FOR AROUND ONE HOUR.

REMOVE FROM HEAT AND LEAVE TO COOL. REMOVE LID
AND CHECK THAT THE BEANS ARE TENDER.

HEAT OIL IN A SAUCEPAN OVER HIGH HEAT.

WHEN IT BEGINS TO SMOKE, ADD THE PORK AND FRY FOR 5
MINUTES ON EACH SIDE UNTIL GOLDEN.

ADD THE BEAN LIQUID, COVER THE SAUCEPAN, BRINGING IT
TO A BOIL AND COOKING IT FOR 10 MINUTES MORE.

ADD THE CHILLI, EPAZOTE, CORIANDER AND SALT AND MIX,
COOK ANOTHER 10 MINUTES UNTIL THE PORK AND BEANS ARE
COOKED AND TENDER.

SALT TO TASTE, REMOVE FROM THE HEAT AND KEEP WARM.
SERVE TWO CUPS OF HOT TAKGSWAYAJUN WITH ONE PORTION
OF PORK IN EACH BOWL.

600 G PORK PIECES
CUT INTO 6 X 100 G SLICES

1/2 MEDIUM WHITE
ONION (100 G)

2 LARGE GARLIC CLOVES,
PEELED AND CUT IN HALF

2 LITRES
WATER

1 LEVEL TABLESPOON
SALT (20 G)

1 LEVEL TEASPOON CHILTEPÍN CHILLI,
ROASTED AND BLENDED (3 G)

1/2 CUP, EPAZOTE
AND CORIANDER
FINELY CHOPPED (26 G)

1/4 CUP VEGETABLE OIL
(60 ML)

South America

FRIJOLES, FRISOLES, FEIJÕES AND MORE

Even though recent studies suggest that beans originated outside the Andean region, it took no time before they were considered wholly South American. This pulse is a staple for millions of people there. Their status is largely due to the bean's historic acclimatisation and early domestication leading it to evolve its very own gene pool in the South American sub-region (like the bean found at the ancient Cueva del Guiterrero site in Peru). Accordingly, there is a startling array of names to describe the pulse: *frijoles, fréjoles, frisoles, feijões, kumandas, porotos, granos* and *caraotas,* to name just a few.

No surprise, then, that the region is home to the International Centre for Tropical Agriculture (CIAT), located in the Colombian Department of Valle del Cauca. This collaborative research body works to improve agricultural productivity and the management of resources of tropical countries, including beans. The Centre houses the largest quantity of germplasm of bean varieties preserved worldwide, and undertakes sophisticated experiments to make these plants more resistant to disease, heat and low-phosphate soil. Researchers strive to increase micronutrients of beans and improve upon the natural nitrogen fixation of pulse roots, which is so highly beneficial to the environment.

Beans are central to the rich cuisines of Peru and Colombia, two countries shaped by countless influences throughout their histories. Indeed, their lexicon is imbued with terms such as *chifa, nikkei, criolla* and *paisa*, referring to dishes influenced by a mix of various cultures. During several decades of the 20th century, different varieties evolved to adapt to the various climate and soil conditions of Peru's coastal region, stretching from Ecuador to Chile. The result is that this area now forms an important production belt spanning some 3 000 kilometres.

In countries like Bolivia, Ecuador, Uruguay and Venezuela, this pulse is a dietary staple due to its high yield, low cost and high nutritional value. The Southern Cone features some important dry pulse-producing regions in the Argentine provinces of Buenos Aires, Córdoba, Jujuy, Salta, Santa Fe, Santiago del Estero and Tucumán. The Puy lentil and the chickpea, grown mainly in Argentina, Chile and Peru, are also common in South America, while other pulses such as butter beans are similarly important to the region.

One country deserves special mention, given its demographic and geographical size: Brazil, which has a longstanding tradition of *feijões* (black beans). This versatile pulse is an ingredient of its national dish, *feijoada*, a unique speciality in the culinary world. Depending on the ingredients, it can go from the most humble, rudimentary dish, to a delicacy fit for a royal banquet. According to texts like Luís da Câmara Cascudo's *História da Alimentação no Brasil*, this is due to the fact that the dish takes its origins from banquets held by plantation owners; leftovers would be collected by slaves and labourers who used them to make a tasty, low-cost meal. The dish is now an institution, as likely to be seen in a luxury hotel as it is in the most humble of households. Its nutritional value stems from a combination of the *feijões*' excellent plant protein, the stewed pork, and high-grade carbohydrates in the form of rice or *farofa* (cassava flour essential to Brazilian cuisine). It can even be combined with vitamin-rich fruit (especially oranges, originally consumed to prevent scurvy) and any other ingredient that the imagination and resources allow.

Another leading Brazilian recipe featuring pulses is *tutu à mineira*, a humble meat and black bean stew made from the leftover broth from *feijoada* or *feijão preto*, thickened with farofa and seasoned with chilli, pepper and garlic. And while chickpeas are fairly rare in the country and seldom grown or consumed, lentils are quite popular in Brazil, though almost all lentils are imported. Brazilians have even adopted the Italian tradition of eating them on New Year's Eve in the belief that they bring prosperity.

Brazil

HELENA RIZZO AND HER INNOVATIVE FEIJOADA

Cosmopolitan flavours and new techniques are central to professional cookery in the world's metropolis. Sometimes, chefs find themselves beholden to the trends and tastes sizzling around an urban environment. São Paulo, for example, has been moving towards more Italian or French ingredients and flavourings. Traditional charcoal grill houses are transformed into buzzing social scenes, keeping in step with the growing status derived from the country's pastures and cattle. While on the home front, beans, cassava and rice remain the main staples, a healthy dietary legacy of colonial Brazil. And while some of the city's most popular restaurants like *A Figueira* still offer a tremendous weekly service of the mainstay *feijoada*, testifying to the lasting importance of this national dish, it would appear that an awkward classist attitude has arisen - one that tends to snub popular, traditional dishes in favour of more trendy tastes, reducing the prominence of locally grown produce and traditionally prepared stews, and diminishing the presence of pulses in some of the best known eateries.

World-famous Brazilian chef, Helena Rizzo, has set out to singlehandedly buck this trend and boldly reinstate the humble pulse. Her signature *feijoada* served up on weekends at her restaurant, *Maní*, restores Brazil's prized dish to its rightful place; positioning it among a host of innovative dishes made using modern techniques. Chefs incorporating progressive cookery with more traditional methods are succeeding in creating some of the finest examples of local cuisine. *Sous-vide* cooking techniques and alginate solutions are part of Rizzo's armoury, while pork ribs and trotters, beef shanks, sausages, chicken, chestnuts, walnuts, orange slices, cabbage and other vegetables, but above all, beans, still feature in her stately and humble version of Brazil's most historic dish.

The results speak for themselves. In 2013, Helena Rizzo was named top chef in Latin America, and the following year, she was crowned the World's Best Female Chef in London, recognised by the same panel of food critics that compiles *The World's 50 Best Restaurants* annually. *Maní* appears on this lofty list at number 41. In 2015, Rizzo garnered a star in the first *Michelin Guide* to Brazil, and her eatery was selected as the city's best contemporary restaurant in the 2015 *Ver São Paulo* lifestyle magazine. *Maní* is situated in the sophisticated district of Jardim Paulistanom, and its name evokes the indigenous goddess of *cassava*, who according to legend, was buried in the place where the most highly prized root in the Brazilian diet first grew.

Born in Rio Grande de Sul, in Porto Alegre in 1978 – the state that borders Uruguay and Argentina and where inhabitants are known as Brazilian *gauchos* – Helena studied architecture during her brief stint at university. At 18, she moved to the seething city of São Paulo, starting work as a model, before becoming involved in cooking with Fasano, one of Brazil's most famous hotel groups. After running the *Na Mata Café* kitchen, she travelled to Europe, where she worked in restaurants in Italy and Spain. At the *Celler de Can Roca* in Girona, Spain (at that time considered the best restaurant in the world), she was introduced to the idea of cooking as an art form. Working alongside Daniel Redondo, whom she met there and married later, she opened her own restaurant, *Maní*, in 2006.

Committed to using fresh, local, seasonal produce and primary flavours, Helena's dishes are in tune with natural cycles, sustainable farming and native Brazilian products. She uses suppliers who carefully select their pulses, like Antonia Padvaiska, of *Emporio Piotara*, who supplies her with an excellent local variety of butter bean – the *manteiguinha del Norte* – or the *Coruputuba* farm, in the Paraiba Valley, her source of cow and pigeon peas. "We are lucky to take advantage of the wide variety of dried beans we have in Brazil," says Rizzo, "Not only the *manteiguinha* and black bean, we also use pigeon peas at *Maní*, a pulse that was widely consumed in the past but fell out of favour. Now, some producers have started replanting it and extolling its virtues, with the result that it has been welcomed enthusiastically by young chefs."

Chef Rizzo selecting pulses at the Pinheiros' market, and in her restaurant *Maní*, in São Paulo.

COWPEAS with CLAMS AND MUSHROOMS

SERVES 4

SOAK
24 HOURS

SOAK BEANS IN WATER FOR 24 HOURS AND DRAIN.

+SALT

250 G COWPEAS *

50 G IBÉRICO HAM

2 BAY LEAVES

BOIL IN FRESH WATER FOR 1 HOUR 40 MINUTES, ADDING SALT TO TASTE, BAY LEAVES AND HAM.

REMOVE BAY LEAVES AND HAM.

BLEND A THIRD OF THE BEANS AND THEIR LIQUID INTO A VERY SMOOTH PURÉE.
MIX BACK IN WITH THE BEANS.

Sauté the mushrooms in a hot pan with a trickle of olive oil and a pinch of salt and add the mushrooms to the stew.

In a medium-sized saucepan, lightly brown the garlic in the remaining olive oil.

Add the clams, white wine and half of the parsley and coriander.

Remove the clams with a spoon as soon as they begin to open.

Before serving, add clams to stew along with salt to taste and the remaining coriander and parsley.

150 G SLICED PORTOBELLO MUSHROOMS

3 GARLIC CLOVES

20 ML EXTRA VIRGIN OLIVE OIL

50 ML DRY WHITE WINE

10 G CHOPPED CORIANDER +10 G CHOPPED PARSLEY

200 G CLAMS

*Scientific name: Vigna unguiculata, also known in Brazil as feijões de corda

FAROFA CAMPEIRA
WITH PIGEON PEAS *

PREPARATION:

1. SOAK BEANS FOR 24 HOURS AND DRAIN.
2. BOIL IN FRESH WATER FOR 1 HOUR 30 MINUTES WITH SALT AND BAY LEAF.
3. REMOVE FROM HEAT AND LEAVE TO COOL AT ROOM TEMPERATURE IN ITS COOKING WATER. WHEN COOLED, DRAIN BEANS AND SET ASIDE.

40 G BUTTER

4 BOILED EGGS CHOPPED

20 ML EXTRA VIRGIN OLIVE OIL

CHILLI OIL

200 G CORNMEAL

HARINA

4. BLANCH THE DICED PUMPKIN IN BOILING WATER FOR A FEW SECONDS AND SET ASIDE.
5. PLACE OLIVE OIL, ONION AND BACON IN A HOT FRYING PAN AND FRY UNTIL ONION IS WELL-BROWNED.
6. ADD THE CORNMEAL AND BUTTER AND STIR TO A CRUMBLY CONSISTENCY.
7. MIX IN THE BEANS AND PUMPKIN AND ADD SALT TO TASTE.

8. SERVE WITH THE CHOPPED EGGS, PARSLEY AND A TRICKLE OF AROMATIC CHILLI OIL.

* SCIENTIFIC NAME: CAJANUS CAJAN

15 G CHOPPED PARSLEY

120 G RED ONION CHOPPED

150 G PIGEON PEAS

1 BAY LEAF

140 G PUMPKIN DICED

200 G STREAKS BACON CHOPPED INTO SMALL CUBES

+ SALT

MY HOMEMADE
BLACK BEAN SOUP

4 SERVES

100 G SMOKED AND DESALTED PORK RIB

50 G ONION

OLIVE OIL

CHILLI OIL

THE SOUP

1 PIG'S TROTTER (PETTITOES)

1. SOAK BEANS IN WATER FOR 24 HOURS.

2. RINSE THE PIG'S TROTTER AND DRY IT USING KITCHEN PAPER AND THEN BURN OFF THE HAIR WITH A BLOWTORCH.

3. DRAIN BEANS AND COOK FOR 1/2 HOUR IN A PRESSURE COOKER WITH THE TROTTER, RIBS, SAUSAGES, "CARNE SECA" DRIED BEEF AND BAY LEAVES. ADD SALT TO TASTE.

4. IN A FRYING PAN, SAUTÉ THE ONION AND GARLIC IN THE OLIVE OIL. ADD TO MEAT STEW, COOKING UNCOVERED UNTIL THE LIQUID THICKENS.

5. STRAIN THE STEW THROUGH A VEGETABLE MILL TO MAKE A SOUP, SALT TO TASTE, SEASON WITH SOME DROPS OF AROMATIC CHILLI OIL AND SET ASIDE.

* BLACK TURTLE PHASEOLUS VULGARIS

2 BAY LEAVES

300 G BLACK BEANS

2 CLOVES GARLIC

50G SKINLESS SAUSAGES

50 G DRIED BEEF, DESALTED

THE CROUTONS

1. MELT BUTTER IN A PAN AND ADD BREAD CUBES.

2. BROWN AND REMOVE THEM FROM THE PAN, PLACING THEM ON KITCHEN PAPER TO ABSORB THE EXCESS FAT.

3. SPRINKLE CROUTONS IN INDIVIDUAL BOWLS, SERVING SOUP ON TOP.

120 G BREAD CUT INTO SMALL CUBES

40 G UNSALTED BUTTER

+ SALT

Europe

THE OLD WORLD'S PANTRY

"**I am unfamiliar with the history of beans on other continents,** but without European beans, the history of those continents would have been different, just as the commercial history of Europe would have been different without Chinese silk and Indian spices", wrote author and semiologist Umberto Eco in a famous article in *The New York Times Magazine*. Displaying both sage judgement and sound evidence, the Italian writer claims that pulses literally saved Europe. He cites that after the long agricultural revolution following the fall of the Roman Empire, the principal source of protein was beans and lentils. These spurred on a process of domestication and storage that helped repopulate the continent during one of its darkest times.

Although people could reap the benefits of pulse protein anywhere in the world, perhaps because pulses were a relative novelty, the Europeans not only embraced them, but made them their own. In fact, looking at traditional recipes across the continent, we find lentils, chickpeas and bean dishes from Norway to Cyprus, from Portugal to Russia, and even in the outermost regions such as the Canaries, the Azores and the French overseas territories. It goes without saying that in certain regions, pulses not only provide a great source of food and nutrition, but are even cultural landmarks. Such is the case of the vast range of pulses from Italy's Emilia-Romagna region, the green lentils of Puy, or the white bean cassoulet of France.

But Europe is also central to the world map of pulses as the point of convergence for the greatest culinary globalisation in history; a process starting with the explorations of

Christopher Columbus and resulting in pulses from Asia meeting their counterparts from the Americas. This gastronomical exchange carried pulses to the four corners of the earth, creating an unprecedented revolution of taste and cultural crossover. Without trade, we would not have seen chickpeas in the Americas and certain kinds of beans in Asia. In the 16th and 17th centuries, the Spanish Empire, the Dutch East India Company (the world's first multinational) and the British and Portuguese crowns were the vehicles that transported these products, perfectly preserved, from agent to agent, colony to colony, and port to port. Works of such universal standing as Miguel de Cervantes' Don Quixote feature a wide variety of dishes, pointing to the rich tradition of pulses by this time, and include references to products within everybody's reach, like chickpeas.

Nowadays, Spain is amongst those countries that are true benchmarks when it comes to pulses; perhaps not in terms of production, where it pales in comparison to countries such as India and Canada, but definitely in terms of diversity and quality produced. There is no other European country that offers such a wide range of pulses, often bearing proud EU labels as a protected designation of origin or a geographically traditional speciality. Indeed, there is no other country where pulses appear so frequently in traditional cuisine.

In Spain, chickpeas are the mainspring of more than 20 different types of *cocido* (a type of stew) served across the country. It is enjoyed from the capital city of Madrid, to Galicia in the northwest, or made in a variant called puchero in the Canaries. Those in the know are particular fans of Castilian interpretations from *Fuentesaúco* (Zamora) and *Pedrosillo* (Salamanca).

As for lentils, *La Armuña* (Salamanca) is home to a delicious variety, with their soft skin and smooth and even texture. They feature in hundreds of traditional recipes and modern creative reinterpretations from the country's top chefs. Spanish beans deserve a full mention too, with some originating in the Middle East and others in the Americas.

Asturian beans (*fabas asturianas*) are one of the leading lights for high-quality Spanish pulses, but equally excellent varieties can be found in *La Bañeza* (León), *Tolosa* (Guipúzcoa) and Segovia, where the *Judión de la Granja* variety was originally animal fodder before it became the love of gourmets. And how could we forget beans from *El Barco* (Ávila), *Ganxet* (Barcelona), *Guernica* (Vizcaya), red kidney beans from *La Rioja* and *Ibeas* (Burgos); each and every one of them representing farming excellence and meticulous methods of production, inextricably linked to each of the regions where they are grown.

MADRID (SPAIN)

Spanish chef Abraham García in his restaurant, *Viridiana*.

Spain

ABRAHAM GARCÍA BRINGS FOSSILS BACK TO LIFE

Glance at the menu at *Viridiana*, Abraham García's restaurant in Madrid, and you will always see pulses amongst the specialities. The humble pulse is not typical fare for top European restaurants, but *Viridiana* has been a quality eatery for 40 years. According to the food critic at the *International New York Times*, it is one of the ten best bistros in the world, a claim backed by Marie Claude Decamps of France's *Le Monde*.

Viridiana's owner and head chef virtually heralded-in multicultural cuisine in Spain; a navigator taking then-conservative Spanish palates on a journey of cosmic fusion to experience a wider world of flavours. Such cutting-edge aspirations strengthened rather than weakened his love for the simplicity of pulses. He researches and creates new tastes with pulses and thereby widens their gastronomic appeal. He presents revolutionary dishes like the Basque black bean stew with okra, the Antarctic stone crab stew, or the New Orleans-style stew with chilli peppers and Cajun spices. These are but a small sample of the creative spirit with which Abraham is infusing pulses, so they can leap out of the past and onto our plates.

Abraham García is a chef who is at once impulsive and sensitive, learned and primal, intuitive and sophisticated – like so many genuinely great chefs. He also has strong beliefs about what food should be, and his daily trips to the markets are translated into his ever-changing menu, which isn't limited to perishable products like fish or vegetables. Abraham brings an unbridled curiosity for new and unexpected tastes, created with countless pulses that have been soaking in water since the night before. He is also the author of several books where literature and food intertwine, such as in *Abraham Boca* and *Recetas para quitarse el sombrero* (Recipes to take your hat off to).

With literary prowess, he quotes the physicist and poet Agustín Fernández Mallo, who believes that "'the reason we love to sit at a table and eat together is because the raw material we pick up at the market is already dead.' Cooking it, serving it and savouring it is the same as bringing it back to life." This brings an awareness of the passing of time, marked by our own inevitable death and magical resurrection.

This tenet is particularly true for pulses. Indeed, Abraham believes that the life cycle of pulses responds to this idea more so than any other product: from its height of freshness as a plant (which implies freezing and thawing for some varieties) and its providential fossil record when dried, to its recovery when soaked, it is given new life in dishes fostering the chef's inspiration and in meals shared by friends. Abraham knows full well that pulses exercise a decisive role in Spain's culinary traditions. From severe stews with *michirones* (which require soaking for 48 hours before an uncertain cooking process); to the subtle taste of

a grass pea flour paste that must be used ever so sparingly (something that people realized only after overeating it in times of famine).

Abraham García was born in the countryside of Robledillo (Toledo), a tiny hamlet in La Mancha. During his bucolic childhood he was a curious rebel. He would go on to train in the great restaurants of Madrid (*Coro, Jockey, Club 31*), but never forgot the basics he learned at home, such as the family recipe called "Three handfuls" (beans, chickpeas and lentils). This dish intuitively foreshadowed the science of creating a healthy range of vegetable protein by serving pulses with a handful of rice thrown in for good measure.

Spain has a range of pulses that proudly bear their own names and seasons, and make both simple and sophisticated regional dishes. Nothing is foreign to Abraham in

the culinary miracle of pulses, the most reliable meal all year-round. He loves *fabada asturiana* (Asturian bean stew) for its immutable rigour; the many types of white beans in Leon, Palencia and Segovia; black and red beans from Tolosa and Guernica, reaching almost holy status when served with local sausages *morcilla de puerro* and *morcilla de repollo*; delicate and moist red kidney beans from La Rioja; and *garrafó* beans that make paella more hearty or add joy to any salad. To this we can add Pardina lentils, Beluga lentils and green lentils, all with their serious character yet so quick to prepare. But, above all, there is the generous chickpea that can do it all – the star of both everyday stews and the stellar version that Abraham only prepares to order. Invariably a one-off, since it is never the same way twice, just as it should be each and every time, for an eternity.

The Spanish chef Abraham García choosing pulses in Casa Ruiz, a specialised shop offering a wide variety of European pulses (top right). The rest of the photos were taken in the chef's restaurant, *Viridiana,* in Madrid.

TAPAS

Tapas are Spain's most popular side dishes. They are an important part of a casual cuisine that frequently accompanies drinking alcoholic beverages.

Tapas can satisfy hunger pangs and accentuate the friendly ambience of bars- where being with friends, and going from place to place eating different tapas, has become a ritual that makes life in Spain different from anywhere else.

Anything is possible and allowed when enjoying tapas, and pulses have their rightful place there, too. A few examples from Chef Abraham García:

"A feira" octopus ON A BED of FRIED CHICKPEAS

8

TAPAS

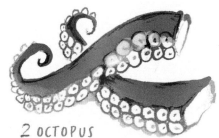

2 OCTOPUS TENTACLES, COOKED AND COOLED

400g CHICKPEAS, COOKED

EXTRA VIRGIN OLIVE OIL

4 GARLIC SCAPES (SHOOTS)

1. FRY CHICKPEAS IN A LITTLE OIL ON MEDIUM HEAT, ADDING THE GARLIC SHOOTS TOWARDS THE END.

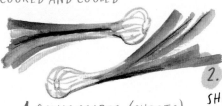

2. SERVE THE CHICKPEAS AND GARLIC SHOOTS ON EIGHT SEPARATE PLATES AND PLACE THIN SLICES OF OCTOPUS ON TOP.

3. SPRINKLE THE PAPRIKA, DRIZZLE A LINE OF OIL AND ADD A LITTLE COARSE SALT.

PAPRIKA 10g

+ COARSE SALT

Chickpea and COD STEW
with
SPINACH and PINE NUT FRITTERS

2 BUNCHES OF SPINACH

1) Place the CHICKPEAS IN WARM WATER and soak for 12 hours. Drain.

2) Place the soaked chickpeas in a pressure cooker, cover them in water, then add the vegetables and a trickle of olive oil.

3) Put the spinach to one side, and then BLEND the other vegetables with some of the cooking liquid.

4) Place the blended vegetables in a saucepan together with the chickpeas and cook on low heat.

OLIVE OIL

1 ONION

500 G DRIED CHICKPEAS

50 G PINE NUTS

12 ALMONDS

BAY LEAF

5 SLICES OF BREAD

2 EGGS

250 G FLAKED COD

+ COARSE SALT

5) <u>FRY</u> EACH SLICE OF BREAD TOGETHER WITH THE ALMONDS AND THE SAFFRON.

6) ONCE FRIED, GRIND THE BREAD, ALMONDS AND SAFFRON WITH A PESTLE AND MORTAR AND ADD IT TO THE SAUCEPAN.

7) DRAIN THE SPINACH, ROLL IT INTO A BALL, AND THEN <u>CHOP IT ROUGHLY</u>.

8) MIX THE SPINACH WITH THE BEATEN EGGS, TOASTED PINE NUTS AND CHOPPED PARSLEY.

9) SHAPE THE SPINACH MIXTURE INTO SMALL BITE-SIZE PIECES THE SIZE OF A SPOON AND FRY THEM IN OIL <u>ON HIGH HEAT UNTIL GOLDEN</u>.

10) COOK THE SPINACH FRITTERS IN THE STEW FOR 30 MINUTES. WHILE ITS COOKING, ADD THE FLAKED COD. IF USING SALTED COD, ENSURE YOU SOAK IT FIRST TO REMOVE THE SALT.

SPRIGS OF FLAT-LEAF PARSLEY

1 LARGE, RIPE TOMATO

1 LEEK

1 G SAFFRON

3 GARLIC CLOVES

1 CARROT

MUSSELS ON A BED OF VERDINA BEANS
with a peach and chilli vinaigrette

PREPARATION:

Steam the mussels and remove one shell. While cooling, prepare vinaigrette made from mixing equal quantities of all of the finely chopped ingredients adding a generous amount of oil, lime juice, Jerez vinegar, and Dijon mustard, Tabasco sauce and coarse salt.

Lime Juice

2 Kg mussels

(STEAMED OPEN AND WITH ONE SHELL REMOVED)

56

PARSLEY

EXTRA VIRGIN
OLIVE OIL

PICKLED GHERKINS

RED ONION

500 G
VERDINA
BEANS

OR

COOKED
HARICOT
BEANS

CAPERS

A FEW DROPS
OF TABASCO

1 TEASPOON
OF DIJON
MUSTARD

A FEW DROPS OF
JEREZ VINEGAR

DRIED PEACH

COARSE
SALT

SCALLOPS WITH Lentils AND SOBRASSADA*

INGREDIENTS

1 CARROT

1 MEDIUM-SIZE POTATO

1 LEEK

+SALT

1 ONION

8 LARGE SCALLOPS

2 GARLIC CLOVES

300 G LENTILS
(IF UNSHELLED, SOAK OVERNIGHT)

PREPARATION:

1 DRAIN LENTILS AND COOK WITH VEGETABLE STOCK IN A PRESSURE COOKER.

2 REMOVE THE VEGETABLES AND MIX WITH SOME OF THE COOKING LIQUID WITH A HAND BLENDER

3 ADD THE SOBRASSADA , MAKING SURE TO BIND THE MIXTURE   SO IT STAYS THICK.

50G SOBRASSADA*
SAUSAGE, SKINNED

4 TASTE AND SEASON, THEN SERVE INTO EIGHT SMALL CLAY BOWLS.

5 STEAM THE SCALLOPS, REMOVE THE MEATY PART AND PLACE ONE SCALLOP IN THE CENTRE OF EACH BOWL, ON TOP OF THE LENTIL AND SOBRASSADA PURÉE.

* SOBRASSADA IS A DELICIOUS, SOFT SAUSAGE MADE IN MAJORCA FROM PORK AND PAPRIKA

Eastern Europe and Western Asia

Lentils are the first pulse described in the Bible, which later also cites the perfect marriage between pulses and cereals. This first reference occurs, appropriately, in the book of *Genesis*. In it, Esau gives up his birthright to Jacob for a bowl of lentil stew. The dish is initially referred to as a "red stew", suggesting that the lentils they ate were usually split and possibly seasoned. The same text also tells the story of the Jews chosen to serve Nebuchadnezzar after his campaign against Jerusalem. Daniel and the

other young men asked the king for water and pulses rather than the other delicacies that were offered to them. Though this odd request at first disconcerted the people responsible for feeding them, after several days, the greater health benefits of the lentils became clear.

The remains of domesticated lentils and chickpeas found in excavations at Hacilar, Turkey, and Jarmo, in Iraqi Kurdistan, led to the conclusion that these crops have been consumed in the area for at least 8 000 years. In the region of Palestine, where people lived in

the contrasting landscapes of arid desert and the rich terrain of the Asi valleys and Jordan, advanced planting and harvesting techniques were developed. Broad beans and lentils played a vital role in bringing about stable settlements, making the region a Phoenician supply point.

The natural wealth of the strip of land between the Tigris and the Euphrates rivers also gave rise to important ancient cities like Babylon. It is widely accepted that the earliest structured writing system was developed in

Mesopotamia around 3200 BC, and evidence suggests that this innovation was a result of the need to control agricultural production and taxes on its output.

Dried lentils, broad beans and chickpeas would have played a vital role in the evolution of this early civilisation due to their long shelf life. Many of the pulses we now know came from this part of the world, unsurprisingly the same region where the first examples of domestication of these crops were found, spread and traded.

Pulses are still a very important part of the region's cuisine. Iraqi and Iranian cuisines (with their roots in ancestral Persian, Assyrian, Babylonian and Sumerian traditions), as well as Arabian, Syrian, Jordanian, Azerbaijani, Armenian and Israeli cuisines (and even the more rudimentary Qatari, Omani and Yemeni cuisines) share many recipes for pulses which differ in little more than their name. This is the case of *fasoulia*, a typical bean stew that can be found anywhere, from Greece to the southernmost part of Arabia.

There is no doubt that the region's tradition of eating a selection of small dishes, widely known by the Persian term *mezze*, has propelled the global reach of various pulse-based dishes made from chickpeas, or lentil salads (*leblebi*, *hummus*, *borani* or *falafel*) and bean purees (*fava*).

Turkey, the bridge between Europe and Asia, is one of the world's leading chickpea and lentil producers and exporters. These pulses have deep roots in Turkey's national cuisine, with Ottoman, Greek, Eastern European, Sephardic and, of course, Middle Eastern influences. Simple and nutritious dishes like rice with chickpeas, known as *nohutlu pilav*, as well as combinations of rice with lentils or beans, are staples in Turkish households, like *mercimek çorbası* (lentil soup) and *tutmaç*, to which noodles are added. Turkey's cuisine also uses plenty of beans, surprisingly perhaps, given that they are far less produced than other pulses. Whether as an accompaniment to meat dishes or in salads, purées, stews or soups – many entirely vegetarian – beans hold a historic place of honour in Turkish cuisine.

ISTANBUL (TURKEY)

Chef Didem Senol in the kitchen of her restaurant, *Lokanta Maya*.

Turkey

DIDEM SENOL'S DEVOTION TO PULSES

Lokanta Maya, which opened in 2010 in the Karaköy district of Istanbul, is one of Turkey's most acclaimed restaurants. Its specialities are inspired by the country's traditional dishes, though at the same time it has a reputation for being at the cutting edge of cuisine. The owner and chef, Didem Senol, went on a formative journey that took her from pursuing a psychology degree at Koç University in Istanbul to enrolling, inspired by the warmth of home-cooked stews, at the French Culinary Institute in Manhattan's SoHo. Before returning home, she gained valuable work experience at the Eleven Madison Park restaurant. Back in Istanbul, she worked at Un Teras, a fashionable urban restaurant specialising in classic ufak yemekler tasting menus (little dishes similar to mezze or tapas). But in 2006, she joined the Dionysos Hotel in Kumlubük, on Turkey's southwest coast, as their Head Chef. Her experience in the Mediterranean kitchen at this unique tourist destination led to her first book, Kızınız Defne'yi oğlumuz Iskorpit (Flavours of the Aegean Sea), and inspired dishes that would become part of her signature repertoire at Lokanta Maya, such as her famous fish ragout with lentils. Didem Senol identifies the concept of originality in the kitchen with origin; or in other words, with the aesthetic and dietary foundations that have been integral to Turkish cuisine since ancient times. Committed to eco-friendly, sustainable produce and the natural seasonal cycles, she uses dried pulses as a trademark of her establishments. This is a recognition both of the likely native origin of some of them, such as chickpeas (a number of researchers ascribe their early use to Turkey), and of the ever-present beans, lentils, broad beans and dried peas in many home-cooked or everyday urban recipes. For instance, at Lokanta Maya hummus is made from chickpeas (standard), with white beans (more rare), and

even with the two pulses mixed together (quite unusual). Pulses are a constant feature in sauces, accompaniments, purées, soups and salads. They are used in starters and mains, from contemporary ragouts of beans, chickpeas and tripe with crushed lemon, to haricot beans with smoked bream, red lentil and bulgur patties, lentil and broad bean soup, to meat stewed with dried peas, white beans with pastrami and many more...

"I love cereals and pulses," says Senol. "I think they're extremely important for the sustainability of food resources, and as a protein supply for the population. We chefs should shed that professional ego that leads to a kind of classism in certain dishes, of using expensive and unnecessary ingredients. We should give up on passing fads and on overly creative endeavours that deny what we have within our reach. We should give pulses credit and give our customers the opportunity to enjoy them, because they are a necessary resource to feed the planet well."

In May 2012, Didem Senol opened *Gram Pera*, an innovative and popular restaurant-bakery concept only serving lunch. Located in the Pera neighbourhood, it offers casual dining that changes with the seasons, and pulses feature in many of its dishes. Her *Gram Maslak*, in Orjin, which opened in 2014, and the *Gram Kanyon*, her most recent endeavour, replicate the experience. In her second book, *Biraz Maya, Biraz Gram* (Some Maya, Some Gram), published in 2014, she hints at the social responsibility that comes with culinary activity, her interest in sustainability and her unbridled devotion to local, seasonal ingredients.

Didem Senol sources produce from all over Turkey, and she is famous for forging friendships with her suppliers. She is particularly interested in working with small producers. "Local ingredients are what define a cuisine," she often says. She visits Istanbul's food markets every day. On Wednesdays, you will find her at *Fatih Pazari*, and on Saturdays she heads to the organic market in Feriköy. There, she buys lentils from Denizli, or the famous *Bayramiç* chickpeas, grown near the Marmara Sea.

Chef Senol buying pulses at the market in Istanbul (bottom center) and in her restaurant *Lokanta Maya*. in the same city.

WARM HUMMUS

250 G DRY CHICKPEAS
(SOAKED OVERNIGHT)

1. BOIL THE CHICKPEAS UNTIL THEY ARE SOFT.
2. DRAIN CHICKPEAS, MAKING SURE TO RETAIN SOME OF THE WATER AND PLACE THEM IN A FOOD PROCESSOR.
3. ADD THE TAHINI, CRUSHED GARLIC, SALT, LEMON JUICE, CUMIN AND SEVEN TABLESPOONS OF THE COOKING LIQUID.
4. MIX IN FOOD PROCESSOR, SLOWLY POURING IN THE OIL WHILE BLENDING.
5. ONCE THE MIXTURE IS THOROUGHLY BLENDED AND SMOOTH, PLACE IT IN A PAN AND WARM UP ON LOW HEAT.

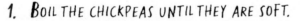

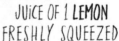

JUICE OF 1 LEMON
FRESHLY SQUEEZED

CAYENNE PEPPER

6. In a separate pan, <u>melt butter</u> with cayenne pepper and add pistachios.
7. To serve, place the warm hummus on a serving dish, season with spicy butter, and drizzle a little oil on top and decorate with <u>pistachios.</u>

2 **GARLIC** CLOVES, CRUSHED

1 TEASPOON OF
SEA SALT

CUMIN

50 G OF BUTTER

50 G OF PISTACHIOS

100 ML OF EXTRA
VIRGIN OLIVE **OIL**

200 **G** TAHINI

250 G OF DRY BROAD BEANS — SOAKED OVERNIGHT

BROAD BEAN DIP

SERVES 4

1. Boil beans until half-cooked (about 20 minutes).

2. Finely chop the onion and sauté it in olive oil.

3. Add onion to broad beans with sugar, salt and 250 ml of boiling water.

4. Cook the broad beans until they are almost a purée.

5. Place mixture in food processor, add lemon juice and gently blend.

6. Serve with the chopped dill, olive oil and fresh spring onions as garnish.

100 ML OLIVE OIL

JUICE OF 1 LEMON

1 ONION

SPRING ONIONS

CHOPPED DILL

1 TEASPOON SUGAR

SALT TO TASTE

RED LENTIL BURGERS

100 ML POMEGRANATE MOLASSES

500 GR OF RED LENTILS

CUMIN

SERVES 4

1 LARGE ONION

FRESH PARSLEY

FRESH MINT

JUICE OF 1/2 LEMON

PAPRIKA

+ SALT
+ 300 ML OF WATER
+ 100 ML OLIVE OIL

1. **P**LACE LENTILS IN A SAUCEPAN AND COVER WITH WATER. SIMMER UNTIL THOROUGHLY COOKED (TENDER) AND ALMOST ALL OF THE WATER HAS BEEN ABSORBED.

2. **A**DD THE BULGUR, STIRRING AND COOKING FOR A FEW MINUTES.

3. **F**INELY CHOP THE ONIONS AND SAUTÉ IN OLIVE OIL, ADDING THE PEPPER PASTE.

30 G OF RED PEPPER PASTE**

250 G FINE BULGUR WHEAT *(OR SEMOLINA)

4. **A**DD ONIONS INTO LENTIL MIXTURE WITH LEMON JUICE, POMEGRANATE MOLASSES, CUMIN, PAPRIKA AND CHOPPED HERBS.

5. **F**ORM SERVINGS INTO PATTIES AND SERVE WITH LETTUCE.

*BULGUR WHEAT CONSISTS OF WHOLE WHEAT GRAINS DRIED IN THE SUN, GROUND AND SIEVED. SEMOLINA IS MADE FROM FINE PROCESSED WHEAT FLOUR TURNED, OFTEN USED TO MAKE COUSCOUS.

** FOR PASTE: TAKE STRIPS OF RED PEPPER AND SLOW-COOK WITH OLIVE OIL, CHILLI, SALT, SUGAR AND WATER.

South and Southeast Asia

THE ORIGIN AND GLORY OF PULSES

American archaeologist and anthropologist Chester Gorman made many notable discoveries in Spirit Cave, Thailand, and his work was continued by his colleague, Wilhelm G. Solheim. Amongst these were finding the remains of domesticated plants, including pulses, dating back to before 9500 BC. Together with other evidence, these findings support theories of farming systems that pre-date even those of the Middle East.

We know that many pulses, some of them extinct today, were domesticated species cultivated on the Indian subcontinent in the Vedic period (c. 1500 – c. 500 BCE). In all probability, these products reached Europe, leading to the cultivation of beans (stemming from the *Vigna* genus rather than *Phaseolus*) well before the discovery of America. It's thought they might have given rise to dishes such as Spanish *fabada* and Occitan *cassoulet*. Meanwhile, lentils (mentioned in the ancient Hindu text *Vishnu Purana*) and chickpeas, originally from western Asia,

acclimatised extraordinarily well and are now found especially in the north of the subcontinent in India, Pakistan and Myanmar (Asia's leading producer of beans).

Southeast Asia is of course home to mung beans (green gram) and mungo beans (black gram), as well as soya beans, all contributing to a unique and varied range of pulse dishes. Countries such as the Philippines, under Spanish influence until the end of the 19th century, or Cambodia, once part of French Indochina, have kept alive certain imported culinary traditions using pulses, even though recipes have been adapted for beans more typical of the region. Elsewhere, since much of the region subscribes to a vegetarian diet and opts for food that is quick to prepare, it is common to find people choosing to eat pulses, raw.

As for the various cuisines scattered across the dozens of islands of the Indonesian archipelago, there are an almost infinite number of uses of dried pulses. The Sundanese in eastern Java use cooked beans such as in the popular dish of *karedok*, while the recipes of West Sulawesi feature pork and beans, with noticeable colonial roots.

In Pakistan, one of the region's largest consumers and producers of chickpeas, we find that dried pulses are the most important source of vegetable protein. In descending order, their main crops include the all-powerful chickpea, mung beans, lentils and mungo beans (black gram), all of which are critical to the country's food security. It should be noted that while many meat products do not fall under *halal* (i.e. not allowed under Islamic law), pulses suffer no such prohibition. Further adding to the prominence of the pulse, the Thal Desert in Pakistani Punjab is an area that suffers from a high level of water stress. Despite being a place hostile to agriculture, it boasts a thriving chickpea crop; significant in that this highlights the power of pulses to grow even in extreme conditions. As for lentils, although there may be fewer than ten varieties, they are the object of future research interest in the country. Lentils are widely accepted in traditional Pakistani cuisine, as can be seen in specialities such as *toor daal*. Of course, an added benefit is that dried pulses are inexpensive. Although they may not normally form part of celebratory meals and banquets, when they do, it is usually to accompany some type of meat dish. Pakistani cuisine also features a number of interesting bean dishes (from both the *Phaseolus* and *Vigna genera*), using tomato sauces with blends of *masala* spices.

India is the world's second largest country by population and is the second largest producer of dried beans, with the greatest amount of land dedicated to their cultivation. It stands to reason, then, that as a country, it also rates highly in terms of overall consumption of dried pulses. India has the highest rate of vegetarianism in the world, with 30-40 percent of its population refusing to eat meat (the percentage is also quite high in neighbouring countries such as Bangladesh and Sri Lanka). Although meat has become increasingly popular in recent years, beans, lentils and chickpeas remain indispensable as a vital protein in the diets of hundreds of millions of inhabitants. This fact has, of course, created an immense culinary heritage orbiting the world of dried pulses.

India

72

SANJEEV KAPOOR, HERALD OF PULSES

It is impossible to understand the world of dried pulses without India. It is the world's largest producer of chickpeas, a leading producer of lentils and beans, and it has a massive population that treats pulses as an everyday staple food. Add to this its rich culinary past and present, as complex as the culture and climate of the states and territories making up the country.

This immense cultural heritage is evidenced in the superb recipes by leading Indian celebrity chef Sanjeev Kapoor, who owns restaurants in virtually all of India's large cities as well as further afield: Bahrain, Bangladesh, Gabon, Jordan, Kuwait, Oman, Qatar, Saudi Arabia, UAE and even Canada. He has made Indian cuisine a household name through his popular *The Yellow Chilli* chain right up to his *haute cuisine* signature brands, *Sanjeev Kapoor* and *Khazana*. In a territory of billions, he has literally carved himself a place in history as one of Asia's most influential chefs. He reaches an audience of untold millions with his Indian cookery programme *Khana Khazana*, which has run for more than 23 years in over 2 000 episodes and is broadcast to over 100 countries.

Multimedia savvy Sanjeev Kapoor is the only chef in the world with his own television channel, *FoodFood*. He has written 200 books translated into seven languages, has millions of followers on social media and his website hosts 15 000 recipes. He takes part in events and TV programmes with famous chefs throughout the world and the Indian government has officially recognised him as the country's top chef. As if that is not enough, he is a social entrepreneur – his companies are committed to gender equality and supporting people with autism. Through his unbridled capacity to communicate, this celebrity chef focuses on finding different ways for people to achieve a healthy and nutritious diet. And through this means, he seeks to prevent disease caused by malnutrition and unhealthy eating habits.

India

"Food should be simple. I believe in combining the best of our traditions, the freshest local ingredients, a bit of science and a dash of art. India is a land of *daals* (*daals* is simply the plural of the Sanskrit word *daal*, referring to dried pulses with the outer hull stripped off, cooked or mashed, or the dishes made with them) and ever since I was a boy, not a day has gone by when they have not been a part of my diet. My favourites are *rājmā* (red kidney beans in a spicy curry) and *pindi chole* (a chickpea curry). Mung and mungo beans (green and black gram) are also very important to my cooking," says Kapoor.

The chef also notes that, given the size of the country, what and how much people eat varies from region to region. Thus, mung and mungo beans are eaten more in the north, lentils and chickpeas in the east, and pigeon peas in the south and west. This fosters a great diversity of dishes which varies region to region, although the philosophy behind them rarely changes. One finds basic dishes from each of these areas, such as *daal tadka* (a curry made with pigeon peas), *sambhar* (a vegetable and pulse curry) and *osaman daal* (a spicy soup made with pigeon peas). Incredibly, even with all this choice, Kapoor says that Indian law dictates that products are clearly labelled either green or brown to distinguish vegetarian and non-vegetarian dishes.

"As a nation we love dried pulses and I am no exception," says Kapoor, who goes on to say that they are "the best comfort food, especially when homemade." Although he loves all pulses, he confesses a particular weakness for chickpeas. He recommends "soaking them overnight and cooking them in a pressure cooker until they are so soft that they melt in your mouth."

The popular chef has successfully used dried pulses in recipes across the board: from salads and soups, to breads, stews and desserts, and he professes that he loves experimenting with them since they "add a unique character to any dish, as well as having the added benefit of their nutritional value." Nowadays, Kapoor is too busy to casually stroll through the markets selecting his vegetables and pulses. But he recommends "A place close to my home, Parle market, in the Mumbai suburbs, where you can find any dried pulse you like." He adds that "Sometimes people are fooled by size when it comes to pulses, thinking larger is better. But in reality, you can often get a lot more taste from smaller grains and beans, and if you have to choose between refined and unrefined pulses, you should always choose the latter."

Chef Sanjeev Kapoor in his restaurant, in Mumbai, India.

TEEN DAL KE DAI BHALLE

SERVES 4

INGREDIENTS:

GREEN CHILLI PASTE
1/2 TABLESPOON

SPLIT GREEN GRAM SKINLESS
(DHULI MOONG DAL) SOAKED
1/2 CUP

PREPARATION

1) DRAIN AND GRIND THE 3 DALS WITH VERY LITTLE WATER TO A FINE PASTE. TRANSFER INTO A BOWL.

2) HEAT SUFFICIENT OIL IN A KADAI (DEEP COOKING POT). WHISK THE BATTER WELL.

3) ADD SALT, GREEN CHILLI PASTE AND RED CHILLI POWDER AND WHISK WELL. DROP SMALL PORTIONS OF THE BATTER INTO HOT OIL AND DEEP FRY TILL GOLDEN.

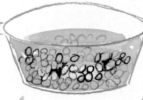

SPLIT GREEN GRAM SKINLESS
(DHULI MOONG DAL) SOAKED
1/2 CUP

SPLIT BENGAL GRAM
(CHANA DAL) SOAKED
2 TABLESPOONS

4) Drain and soak in water for 10 minutes. Drain the bhalle and squeeze out extra water and arrange them on a serving platter.

OIL TO DEEP FRY

Red **CHILLI** POWDER
1/2 TEASPOON

5) Pour chilled yoghurt over them and serve sprinkled with black salt, red chilli powder, cumin powder, date and tamarind chutney and coriander leaves.

+ **SALT** TO TASTE

BLACK **SALT** (KALA NAMAK)
1/2 TEASPOON

6) Serve chilled.

ROASTED CUMIN POWDER
1 TEASPOON

YOGHURT WHISKED AND CHILLED
2 1/2 CUPS

SWEET DATE AND TAMARIND **CHUTNEY**
1/2 CUP

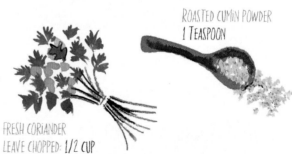

FRESH CORIANDER LEAVE CHOPPED: 1/2 CUP

Qvaabooli MAIN COURSE

RICE: 1-1/4 CUPS

LEMON JUICE: 2 TABLESPOONS

CUP PEELED CHICKPEAS (CHANA DAL): 1/2 CUP

1/2 TEASPOON POWDERED TURMERIC

4 PEOPLE

5 ONIONS, SLICED INTO ROUNDS AND BROWNED

GREEN CHILLIS CHOPPED: 2-3

+ **SALT TO TASTE**

OIL: 2 TABLESPOONS

1) SOAK RICE AND CHICKPEAS SEPARATELY FOR 30 MINUTES. BOIL RICE IN 3 CUPS OF SALTED WATER WITH HALF OF THE WHOLE GREEN CARDAMOMS, CINNAMON AND CLOVES, STOPPING BEFORE FULLY COOKED. REMOVE FROM HEAT AND DRAIN ANY EXCESS WATER.

2) BOIL CHICKPEAS IN 1 CUP OF SALTED WATER WITH HALF OF THE TURMERIC POWDER UNTIL JUST COOKED. DISSOLVE THE SAFFRON IN THE MILK AND SET ASIDE.

3) HEAT OIL IN PAN AND ADD THE REMAINING GREEN CARDAMOMS, CINNAMON AND CLOVES AND SAUTÉ. ADD THE CARAWAY SEEDS AND WHEN THEY CRACKLE ADD THE GINGER AND GARLIC PASTE AND SAUTÉ. NEXT ADD THE CHOPPED GREEN CHILLIES AND A FEW OF THE ONIONS AND BROWN.

4) ADD TO THE COOKED CHICKPEAS THE GARAM MASALA AND THE FINELY CHOPPED CORIANDER LEAVES. STIR TO MIX WELL. ADD THE TURMERIC POWDER, STIR AND REMOVE THE MIXTURE FROM THE HEAT. ADD THE YOGHURT AND MIX WELL.

5) TRANSFER HALF OF THE CHICKPEA MIXTURE INTO ANOTHER FRYING PAN. SPREAD HALF THE RICE OVER THE CHICKPEA MIX, THEN ON TOP SPRINKLE HALF OF THE REMAINING BROWNED ONIONS, TORN MINT LEAVES, LEMON JUICE AND SAFFRON MILK.

6) SPREAD THE REMAINING CHICKPEA MIXTURE OVER THE RICE, FOLLOWED BY THE REST OF THE BROWNED ONIONS, GARAM MASALA POWDER, TORN MINT LEAVES, LEMON JUICE AND SAFFRON. COOK FOR 20 TO 25 MINUTES (COVERED TIGHTLY WITH A LID AND EDGES SEALED WITH DOUGH TO PREVENT THE STEAM FROM ESCAPING) AND SERVE HOT.

1 TEASPOON POWDERED GARAM MASALA

YOGHURT 1/2 CUP

CARAWAS SEEDS (SHAHI JEERA): 1/2 TEASPOON

A PINCH SAFFRON

FRESH CORIANDER LEAVES CHOPPED: 1 TABLESPOON

GINGER AND GARLIC PASTE: 1/2 TEASPOON

CLOVES: 4-5

MILK: 2 TABLESPOONS

5CM CINNAMON STICKS

FRESH MINT LEAVES, TORN FROM A FEW SPRIGS

4-5 GREEN CARDAMOMS

Madgane

DESSERT

1) Cook chana dal and cashewnuts with 1½ cups water in a deep non stick pan till just done. You can even pressure cook them. Cook jaggery with 1 cup water till it dissolves. Mix rice flour with ¼ cup water to a smooth paste.

Split Bengal gram (chana dal) soaked: ½ cup

Green Cardamom Powder :
½ Teaspoon

Thick Coconut Milk :
1 cup

Jaggery (Gur)
Grated: 1 1/2 cups

2) Add thin coconut milk to the jaggery water and mix well. Add the rice flour mixture and cook on medium heat, stirring continuously. Add green cardamom powder and cook till the flour gets cooked and the mixture thickens.

3) Reduce heat and add thick coconut milk. Mix well and switch off heat. Serve warm.

Thin Coconut Milk:
1 1/2 cups

Cashewnuts
1/2 cup

Rice Flour :
3 tablespoons

KARACHI (PAKISTAN)

Chez Zubaida Tariq at her home, where she holds cooking and film classes and does catering.

Pakistan

ZUBAIDA TARIQ, REVEALING WISDOM

Pakistan is the world's third largest producer of chickpeas, and it is also a significant consumer of lentils. The words *chana* (chickpea) and *daal* (lentil) recur in Pakistani cuisine, and yet agronomists are always hard at work, looking for ways to produce even bigger and better harvests of pulses.

The country has produced many highly respected chefs, but Zubaida Tariq (born in Hyderabad Deccan in 1945) deserves a special place among them. As a TV chef and authority on food, she has been followed faithfully by millions of viewers for over 20 years. Although born in India, in the region currently comprising the states of Andhra Pradesh, Karnataka and Maharastra, she grew up in an affluent family of Urdu speakers that has produced a bevy of talented writers, artists and intellectuals. Her siblings are deceased novelist and playwright Fatima Surayya Bajia, poet Zehra Nigah, fashion designer Sughra Kazmi and the renowned television host and humourist Anwar Maqsood Hameedi. The first years of Zubaida's life coincided with the end of British rule on the subcontinent, and she moved to Pakistan with her family in 1947, the year it gained independence following the partition of India.

Known affectionately as Zubaida Apa (meaning "big sister" in Urdu), she came to cooking relatively late in life. However, when she did, she bloomed, to the point that she is now an undeniable reference point for cooking across the region. This is largely due to her numerous appearances on radio and television, such as the popular show *Handi*, on the food channel *Masala TV*. "When I married at 21, I couldn't cook a single dish. My mother, who was good at many things, had taught us to push ourselves to achieve whatever it is we desired, but she didn't teach me to cook. So once I decided I wanted to cook well, I took her advice and taught myself, by trial and error. Little by little, I realised that a lot of people liked what I cooked," she says.

Pakistan

Zubaida Apa has captivated audiences with her personal style of cooking, based on traditional Pakistani cuisine and prepared with simple ingredients, such as the country's ever-present pulses. Her popular recipe books, including her take on dishes such as the national favourite *daal masala*, are highly influential in home cooking throughout Pakistan. Through her tried and true recipes, regional specialities have made their way around the world. "Dishes such as *chana chaat* and *murgh cholay* are popular all year round. *Lobias* (black-eyed peas) are excellent together with a light tamarind paste or a fresh tomato sauce, whether with a *chapati* (a type of wholewheat flatbread) or as a dressing. Lima beans can be eaten fried, Chinese style. *Masoor* (pink lentils) and mung beans (green gram) can be stewed with spices and then mixed with rice or curry. *Daal mash* is very popular, but it is more of a celebration food." Amongst her favourite dishes are, of course, pulses. "At home I like to eat *keema* (a minced mutton curry with peas) with rice, *papri chaat* and *daal*," she confesses. "In fact, *daal* is a food that should always be on the dining table: it is healthy, cheap, nutritious and tasty."

This popular TV personality has won over the entire country with both her cooking and beauty tips, and she is convinced that "plant-based ingredients make a kitchen come alive." Although she says that it is easy to find pulses in Pakistan both in supermarkets and from street vendors, she prefers to go to the Empress Market in Karachi, "first thing in the morning, when the traders are setting up their stalls and goods are only just arriving. In just one trip I know I can get all of the ingredients I need, at the best price and also the best quality."

"Pakistani pulses have taken on their own personality in the food of the subcontinent, but we need to have a wider range of vegetarian dishes and eat far less meat in our diet. We can do creative things with pulses, with all manner of ingredients. We can boil them, stew them, layer them, turn them into fritters, vegetable kebabs, and so much more."

Chef Zubaida Tariq in the Empress market in the centre of Karachi, Pakistan, and in her home, preparing and mixing ingredients for a recipe with pulses.

Punjmel Lentils

Pink LENTILS,
YELLOW LENTILS
1/2 CUP EACH

PIGEON PEAS 1 TBSP

BRAISING

1 ONION (FINELS SLICED)

6 DRIED, ROUND RED CHILIES

GARLIC (4 CLOVES FINELS CHOPPED)

CUMIN SEEDS (1 TSP)

OIL (1/4 CUP)

GINGER /
GARLIC PASTE
1 TBSP

TURMERIC POWDER
1 TSP

HOT WATER
2 CUPS

GINGER
(FINELY CUT)
2 TBSP

RED CHILI POWDER
1 TSP

- SOAK ALL LENTILS FOR 2 HOURS
- THEN BOIL WITH TURMERIC POWDER,
RED CHILLI POWDER AND GINGER / GARLIC PASTE

- **B**LEND WITH HAND BLENDER WHEN LENTILS ARE TENDER

  - **A**DD DRIED MANGO, GREEN CHILIES, BUTTER AND SALT
  - **A**DD WATER, RAW MANGO AND GINGER AND LET IT COOK
  - **F**RY TEMPERING INGREDIENTS IN A FRYING PAN
  - **P**OUR TEMPERING OVER LENTILS AND SERVE.

GREEN CHILIES
(FINELY CUT)
4 PCS

GRAM LENTILS AND WHITE LENTILS
1 TBSP EACH

BUTTER
2 TBSP

1 SMALL RAW MANGO
(FINELY CUT)

DRIED MANGO
6 PCS

SALT
TO TASTE

Yellow Lentils with SOYA

- Drain the soaked lentils and bring them to boil in 2-3 cups of fresh water
- Blend with a hand blender once lentils are tender
- Drain the chopped soya and add to the lentils
- Add salt, green chilies, dried red chilies, turmeric, butter and let it cook with some added water
- In a separate pan dried, fry garlic in some oil and butter and pour over the cooked lentil soya
- Serve hot with chappati.

SERVES 4

TURMERIC POWDER
1 TSP

BUTTER
2 TBSP

COOKING OIL
1/2 CUP

6 TO 8 DRIED
ROUND RED PEPPER

3 GREEN CHILIES
FINELY CHOPPED

YELLOW LENTILS
1 CUP, WASH WITH LUKE
WARM WATER AND SOAK

SOYA
2 BUNCHES...
CHOP AND SOAK IN WATER
WITH 1 TSP TURMERIC

4 GARLIC CLOVES PEELED
TO BE FINELY CHOPPED

SALT TO TASTE

SPINACH and WHITE LENTILS

SPINACH
1 KG

4 GREEN CHILIES

4 GARLIC CLOVES

HOW TO PREPARE:

BUTTER
2 TBSP

WHITE
LENTILS
1 CUP

- WASH THE LENTILS WITH LUKE WARM WATER AND SOAK FOR A WHILE
- CHOP AND WASH THE SPINACH
- MIX SPINACH WITH THE LENTIL AND LET IT COOK ON LOW FLAME
- ONCE THE LENTIL AND SPINACH IS TENDER, BLEND WITH A HAND BLENDER
- PUT SALT, RED PEPPER, FINELY SLICED GINGER AND SOME BUTTER AND

COOK FOR ANOTHER 10 MINUTES

- HEAT THE OIL WITH SOME BUTTER IN A FRYING PAN,

AND FRY THE GARLIC CLOVES GOLDEN BROWN.

- POUR IT OVER THE COOKED LENTILS
- GARNISH WITH GREEN CHILIES AND SERVE WITH HOT CHAPPATI.

GINGER
(FINELY SLICED)
2 TBSP

COOKING OIL
1 CUP

SALT
TO TASTE

10 DRIED, ROUND RED PEPPER

87

The Far East and the Pacific

In some countries of the Far East, dried pulses like chickpeas, lentils and beans of the genus *Phaseolus* are largely regarded as exotic, at least as far as everyday domestic use is concerned. This mainly stems from a subsistence farming tradition that since ancient times has revolved around wheat, rice, millet, sorghum and a single pulse, the soya bean, which has an immense prominence in several countries of the region like China, Japan, Taiwan, and the Republic of Korea (South Korea). Another pulse of the genus *Vigna*, the mung bean, is also prevalent, having spread throughout the region after originating in Asia.

But this does not mean that chickpeas, lentils or *Phaseolus* beans are not produced or consumed in the Far East and Oceania. China is one of the world's leading bean producers, while Australia is a major lentil and chickpea grower. The varied weather conditions of this vast region – even with violent phenomena like monsoons – and its immense farms have enabled competitive pulse cultivation.

Japan, the Republic of Korea and Taiwan consume large quantities of dried mung beans and black gram (also of the genus *Vigna*), used in a wealth of traditional dishes. This may in part be due to the influence of western cultures that have left their mark since the early contact between civilisations, from the long colonial period to the recent era of globalisation.

However, there are also many varieties of dried pulses that are eaten locally in certain countries or regions, such as the *Vigna umbellata* or ricebean, originating in Papua New Guinea and consumed in southern China and Nepal. Others are only used for specific recipes such as certain desserts, like the sweet *adzuki* bean originating in East Asia (originally cultivated in the Himalayas), which is widely used in Japan and China to make red bean paste.

Nepalese cuisine, meanwhile, has the traditional *kwa ti* soup, eaten during the *Gun Punhi* festival and made from no fewer than nine types of pulse; some dried and others fresh, including chickpeas, black-eyed peas, black gram and mung beans. This mountainous country is also a major producer of lentils, which are often used in *daal bhat* along with rice, one of the most widespread

culinary combinations across much of Asia. In China's Uighur regions, Mongolia and eastern Russia, pulses are not commonly consumed, but occasionally appear in certain dishes, such as the Russian salad vinaigrette, with white beans, or in Mongolian *kashk*, which is sometimes made with mung beans.

In Oceania, and particularly across the many Pacific Islands, there is no great tradition of eating pulses, although nowadays, their consumption has increased. Australia and New Zealand have both moved toward becoming major producers responding to the influence of global gastronomic trends.

As for China, the world's most populous country and the third largest by land mass, pulse growing has always centred around the soybean, a product that was rare in the West until the second half of the

twentieth century, despite having arrived in Europe in the seventeenth century. As mentioned in the records of the Emperor Sheng Nung, the Chinese were early Neolithic farmers, developing highly advanced agriculture systems in later periods.

Chinese cuisine is as complex and as vast as the country's territory. Today, global trends toward fusion food and the cosmopolitan nature of cities like Hong Kong, Macau and Shanghai, have led to a cultural shift towards pulses in the region. China's major cities have started to adopt a variety of international culinary habits and trends making it easy to find pulses, such as those seen on the menus in increasingly popular Spanish and Mexican restaurants. Despite China's large-scale production of lentils and beans of the genus *Phaseolus*, these crops are primarily grown for export and it is less easy to find them in street markets as one might expect.

BEIJING (CHINA)

The Chinese chef
She Zengtai at his
home with a plate of
bean roulade.

国际烹饪艺术大师
Chinese culinary master

90

China

SHE ZENGTAI, TIME TRAVELLER

The culinary history of the family of the honourable chef She Zengtai, born in Beijing in 1955, dates back to 1404, during the Ming Dynasty reign of the Yongle Emperor. Twenty-two generations have passed on their knowledge, uninterrupted. "Many recipes that I have inherited from my ancestors," he explains, "are over a century old, and some are several hundred years old. The dishes, of course, have been perfected through exploration and the creativity of my ancestors." The fact that this culinary heritage has survived through time is in itself a credit to chef She. After all, preservation of the family trade is part of the complex and ancient history of Chinese cuisine, a tradition spanning over 5 000 years with huge global influence. Chef She has been appointed by the Chinese Government to be the custodian of this intangible cultural heritage.

She, now retired from the everyday activity of professional kitchens, but a fixture on cookery programmes, is a living encyclopaedia of Chinese cuisine. He is able to reel off the uses and customs of each region of this immense country one by one, describing the reasons for each technique, habit or ingredient. Throughout his career, he has prepared countless banquets for political and business leaders, but he has never forgotten his bond with the countryside or his calling to serve the people. Before learning the ins and outs of the profession from his father, She Chonglu, a grand master of Hui cuisine, he worked on the Xiaojialin farm unit as an engine driver.

"As society develops and quality of life improves," says chef She, "people begin to attach more importance to health. As cooks, we must keep up with the times by creating healthy, nutritious dishes and sharing them with society. We have to adjust our cooking methods, like using high temperatures, which can destroy many nutrients, but also the contents of dishes. Experience has shown us that replacing a certain amount of animal produce with dried pulses is the best way to solve the two-way, paradoxical problem of global malnutrition and over-nutrition."

Despite the predominance of soybeans in China, She explains that the volume of production of other pulses is also quite high. Once, when involved in an exhibition on culinary history organised by his country's government, many of the traditional ingredients shown were dried pulses. This provoked surprise among the younger generation of chefs yet received praise from the Chinese culinary community. "They are now iconic products that we Chinese chefs can take abroad with pride, as a manifestation of what future global cuisine must be. Excluding dried pulses from our diet is a serious mistake and it leads to a terrible nutritional imbalance. Dried pulses help us reduce fat levels and strengthen our immune system. They contain protein, amino acids, carbohydrates, B vitamins, carotene and inorganic salts like calcium, phosphorous, iron, potassium and magnesium, but they have a low sodium content," says chef She, who, in addition to being a renowned cook is an expert in traditional Chinese medicinal foods.

She Zengtai is an innovator who has reconciled the rituals of *Hui* and *halal* cuisine, investigating their dietary taboos while finding creative new ways to achieve ethnic culinary perfection. "I like using soybeans, peas, red *adzuki* beans and mung beans. All of these pulses are very versatile. I use them in 20 percent of my cooking, married with meat and vegetables, as well as tofu. To cook them, I like to use a pressure cooker, but then, depending on the dish, other methods can be used," he explains.

As for buying pulses in his area, She explains, "I get pulses in Beijing from places like the great Niujie Muslim Market, where there is a strong affinity with these products, sometimes at the Lotte Mart supermarket, and also in the Dongcheng district. You won't find much difference in the price of the products, but the important thing is to make sure of the quality. Every time I go, I meticulously select the pulses and do a lot of comparing before buying. A recipe should always be tackled using the right ingredients."

Chef She Zengtai buying ingredients in a local market in Beijing and cooking pulses at home.

Yellow SPLiT PEa Pudding

INGREDIENTS:

500G DRY PEAS
SOAKED OVERNIGHT

200G WHITE SUGAR

1 WASH THE PEAS AND PLACE THEM INTO A POT.
ADD FOUR TIMES AS MUCH WATER AND BRING TO A BOIL.

2   Reduce the heat and cook until the peas split.

3   Drain using a colander.

4   Crush the peas to a mash and add the white sugar, mixing well until the sugar has dissolved.

5   Pour the mixture into a pudding form tray.

6   Wait until the mixture has completely cooled before turning it out onto a chopping board, cutting into block and transferring to plates.

HALLMARKS:

The pudding's yellow color and the rich, pleasing flavor of the peas made this dish a wellknown "royal favorite" of the Qing Dynasty.

BEAN ROLADE

1 Knead the cooked glutinous rice flour until it makes a long roll.

2 Cover a chopping block with the cooked soy flour. Then, using a rolling pin, roll out the cylinder into rectangular pieces roughly 4 mm thick.

3 Spread the red bean paste evenly over the pieces.

4 Starting from the end, roll these pieces together gradually until you have a long cylinder three centimeters in diameter.

5 Cut into pieces and stack them onto a plate.

HALLMARKS:

This roulade has a smooth, delicate texture and a sweet flavor and was another favorite of the royals during the **Qing Dynasty**.

500 G GLUTENOUS RICE FLOUR, COOKED (20 MINUTES)

150 G RED BEAN PASTE

100 G **SOY FLOUR**, COOKED (1/2 HOUR IN PRESSURE COOKER)

Lily Broad Beans

1   Heat a frying pan on high heat and pour in cooking oil.
2   Sauté onion and ginger before adding the broad beans and stir-frying.
3   Add the lilies, red pepper, sugar, vinegar and salt.
4   Add starch combined with a little water. Stir-fry and serve.

HALLMARKS:

This dish contains an array of colors, with red, green and white. Fresh and delicate, crunchy yet soft, with a light and refreshing flavor that makes for a modern, healthy dish.

250 G FRESH BROAD BEANS, COOKED

50 G FRESH LILIES

50 G COOKING OIL

25 G RED PEPPER

ONION, GINGER, SUGAR, WHITE VINEGAR, SALT, A LITTLE CORN STARCH

North Africa

There are practically as many pulses in North Africa as there are sand dunes in the Sahara, but it is the chickpea that stands alone. In all probability, they arrived in the region in ancient times when the Phoenicians brought with them numerous vegetable species.

From the legendary city of Tyre, the Phoenicians founded the Punic State, with Carthage as its capital in what is now Tunisia. From there, the chickpea migrated to Western Europe via the Iberian Peninsula, before spreading across most of the Mediterranean during various military campaigns. While the Islamic conquest of the Maghreb brought still other pulses to the region, it reaffirmed the chickpea's predominance as a food staple.

Coming from a hostile geographical context in which inhabitants have always had to outsmart the desert, the dry chickpea has steadfastly remained a reliable food source for the region's nomadic tribes. Thanks in part to its unquestionable nutritional value and long shelf life, the diplomatic pulse helped eliminate boundaries by sharing a culinary tradition across the continent – from the Western Sahara to Alexandria.

In this region, Berber cuisine offers highly nutritious dishes such as *chakhchoukha*, a mixture of *rougag* bread and *marqa*, a lamb, vegetable and chickpea stew. While these recipes find their roots in Algerian cuisine, they take on variations throughout the area, such as *lablabi*, a Tunisian soup.

It is virtually impossible to pinpoint the geographical place of birth of the domesticated pulses of North Africa – most likely it occurred in various places at once. What we can be certain about is that along the Nile Valley, long before the pharaohs, in Neolithic times, the soil fertilised with lime from the river and played a decisive role in leading to the emergence of one of the most fascinating civilisations in human history. In Egyptian hieroglyphics, a symbol resembling three dots is a distinct reference to seeds, quite possibly pulses, which were the main source of protein for these inhabitants of ancient Egypt. They subsisted primarily on chickpeas, broad beans and lentils. In the *Dra' Abu el-Naga'* necropolis near Thebes, funeral offerings in the form of lentil cakes have been found, and it is widely accepted that lentils were one of the most highly prized pulses. Frescoes from the Nineteenth Dynasty of Egypt show a servant cooking these pulses, while the city of *Phacusa* was known as the "City of lentils".

Algeria, Egypt, Libya, Morocco and Tunisia are countries where pulses play a central role in their cuisines (with Sudan, perhaps, the only regional exception). In addition to a variety of chickpea dishes such as *hummus* and *falafel*, lentils are also common, in salads, purées and stews. Significant, too, is the nutritional value they provide people during the month of *Ramadan*, when eating is forbidden during daylight hours.

In Morocco, pulses are especially prominent in dishes such as *tajine*, where chickpeas almost always feature. In the traditional *harira* soup, in addition to chickpeas, dried peas and lentils add further substance, particularly during the winter months. Pulses also feature in the hearty broad bean purée, *bissara*, a mainstay of communities in the Atlas Mountains. In Moroccan markets, street vendors sell cornets of steamed chickpeas dusted with cumin, in the same way that chips are sold at fairs in Western countries. Another culinary feature are kiosks where cooks prepare *kalinti*, or *karane*, using chickpea flour, a speciality of Sephardic origin. It is served in rolls or paper bags as a fortifying snack for workers or school children. Also for Moroccans, lentils are standard fare in stews and salads prepared in households and for important celebrations.

Morocco

MOHA AND THE JEWELS OF TAJINE

The Swiss *Vieux-Bois* restaurant, part of the École Hotelière Genève, is located opposite the United Nations headquarters in Geneva. Due to its proximity, culinary integrity and cosmopolitan service, with students from over 40 countries, its dining rooms host a mainly international clientele. It must be one of the most efficient "hands-on" hospitality schools in Europe, with restaurants and gardens open to the public. Its versatile students, studying cookery, hospitality and restaurant management are trained to become both chefs and managers.

Mohamed Fedal came to the *Vieux-Bois* for three years at the age of 18, (1985-1988). He remained in Europe where a culinary revolution was underway in the early 1990s before gaining experience in the kitchens of famous North American hotels. Ultimately, homesickness brought him back to his native Marrakesh. Mohamed Fedal, known in professional circles as Moha, now 30, purchased the home of French fashion designer Pierre Balmain, located in *Dar El Bacha*, his childhood neighbourhood, near the Medina of Marrakesh. And in 1999, he opened the *Dar Moha* Restaurant, where he returned to the fundamentals of traditional Moroccan cuisine. While adding some innovative touches to his dishes, he set out to earn Intangible

Cultural Heritage status for his country's culinary traditions.

It is in the flavours harking back to his childhood where Moha's taste for pulses lies. Along with couscous, argan oil and the indispensable *ras el hanout*, that combines up to 40 spices, they form the basis of his inspiration as a chef. It's a way of cooking learned at his mother's side, a master chef in a family of painters and artisans. "As a boy, I learned that pulses are the jewels of Maghreb cuisine. Because of their variety and colour – dried beans and peas, chickpeas or lentils; round, flat or long; green, red, yellow or black – they nourish and crown a *tajine* like diamonds, rubies or emeralds."

Morocco

Moha's enthusiasm for pulses is not purely a romantic notion. Some six kilometres from the centre of Marrakesh, he runs the *Riad Le Bled*, a rural hotel on a three-hectare estate. There, he provides accommodation and dining areas for banquets and celebrations. But first and foremost, the orchards, vegetable gardens and the pulses that are grown and dried there are used to supply his restaurants with the beloved chickpeas, green beans, broad beans, peas and lentils that feature heavily in his dishes.

Moha Fedal is now leading a revolution in Moroccan cuisine, a challenge that he tackles by sharing his vision and know-how, and providing practical training in the kitchen to future professionals, including schoolchildren; educating them on food and culinary self-sufficiency. As a professional and teacher, he asks himself a question: "If choosing forces you to give something up, why choose between tradition and modernity when you can combine the two trends harmoniously?"

In his quest for a lighter cuisine, he refines quantities and adds nuances. Under his guiding hand, revamped *hariras*, hummus and pulse salads take on a new form without succumbing to products or condiments from outside Morocco's traditional culinary culture. Doing so would globalise them and strip them of their identity. At the same time and, ironically, his pulse recipes have achieved an even broader appeal.

Moha Fedal is a winner of the Vermeil Medal of the French Academy of Arts, Sciences and Letters, an accolade recognising understanding and dialogue between cultures and societies. He is the author of cookbooks, such as *The Flavour of Morocco, Moha's Kitchen*, and others, all of them published in France where he is highly regarded as a top chef. He is often Morocco's official representative at international culinary exhibitions, a Moroccan Master Chef judge and, every Sunday at 10 am, he participates in the radio programme *Family Kitchen*.

In 2015, Moha brought his carefully orchestrated updates of traditional Moroccan culture to the World's Fair in Milan. There, he presented Moroccan cuisine at a universal stage that focused on soil, nutrition and the right to a healthy and adequate diet. Naturally, pulses played a starring role.

Chef Fedal buying pulses in the market, Mellah in Marrakesh and preparing a dish with pulses in the kitchen of his restaurant, *Le Bled.*

HARIRA

150 G LENTILS

150 G CHICKPEAS

150 G DRY BROAD BEANS

OLIVE OIL

2 TABLESPOONS

APPROX. 3 LITRES WATER

1 GRATED ONION

GINGER GRATED TO TASTE

SERVES 6

PREPARATION:

1. SOAK THE DRY BROAD BEANS, CHICKPEAS AND LENTILS FOR EIGHT HOURS.

2. DRAIN AND PLACE THEM IN A SAUCEPAN WITH GRATED ONION, PEPPER, GINGER, TURMERIC, CINNAMON, CORIANDER, CELERY AND OIL.

3. COVER WITH HALF THE WATER AND BRING TO A BOIL.

1/2 STICK
CELERY,
CHOPPED

4 SPRIGS
FRESH
CORIANDER

1 STICK
CINNAMON

TURMERIC
GRATED
TO TASTE

500 G CHOPPED
TOMATOES

TOMATO
CONCENTRATE

2 TABLESPOONS

4. REDUCE HEAT AND SIMMER FOR 15 MINUTES.

5. ADD THE CHOPPED TOMATOES AND TOMATO CONCENTRATE AND THEN ADD ENOUGH WATER TO COVER WELL AND COOK FOR ANOTHER 30 MINUTES.

6. MIX FLOUR WITH ABOUT 2 CUPS OF WATER TO MAKE A SMOOTH MIXTURE WITHOUT LUMPS. POUR SLOWLY INTO THE POT, STIRRING CONTINUOUSLY WHILE HEATING FOR ANOTHER 20 MINUTES.

7. SEASON WITH SALT, LEAVE TO STAND AND SERVE OVER RICE OR NOODLES.

150 G FLOUR

50 G NOODLES

50 G RICE

1 TEASPOON
PEPPER

SALT
TO TASTE

KORAIN  (Moroccan Hargma)

6 CLOVES
GARLIC
CHOPPED

2 LITRES
WATER

GROUND
GINGER 1 TABLESPOON

2 CALF'S FEET
(OR SHEEP) SLICED

250 G
CHICKPEAS
(SOAKED FOR 8 HRS)

CUMIN
1 TABLESPOON

PAPRIKA
1 TABLESPOON

PREPARATION:

1. PLACE THE SLICED FEET IN A LARGE COOKING POT WITH CHICKPEAS, GARLIC, PAPRIKA, CUMIN, GINGER AND OLIVE OIL.

2. SAUTÉ OVER MEDIUM HEAT FOR 15 MINUTES.

3. ADD THE WATER AND BRING TO A BOIL.

OLIVE
OIL 8 TABLESPOONS

+ SALT
TO TASTE

4. THEN REDUCE HEAT AND ALLOW TO GENTLY SIMMER FOR APPROX. 4 HOURS.

5. TASTE AND SEASON WITH SALT.

HARGMA IS ANOTHER TRADITIONAL NORTH AFRICAN DISH. IT IS A SAUCE MADE WITH GRILLED SHEEP OR CALF'S FEET, AND IS USUALLY EATEN WITH BREAD.

HUMMUS

PREPARATION:

1 SOAK THE CHICKPEAS IN A LARGE CONTAINER COVERED IN PLENTY OF WATER FOR 6 HOURS. THE CHICKPEAS WILL SWELL. DRAIN AND PLACE THEM IN A COOKING POT, ADD WATER AND BRING TO A BOIL.

2 LOWER THE HEAT AND SIMMER COVERED, ADDING MORE WATER IF NECESSARY, AND COOK FOR 1-1/2 HOURS.

CHICKPEAS 450g
COOKED AND DRAINED

3 DRAIN CHICKPEAS KEEPING THE COOKING WATER, AND THEN PUT THEM ALONG WITH A SMALL AMOUNT OF COOKING WATER IN A BLENDER.

4 ADD THE TAHINI, GARLIC, CUMIN AND LEMON JUICE.

5 WHILE BLENDING, ADD THE OLIVE OIL TO MAKE A SMOOTH, VELVETY CONSISTENCY, THEN SEASON WITH SALT.

6 SPRINKLE WITH CUMIN, PAPRIKA AND PEPPER TO TASTE. SESAME OIL AND OLIVE OIL ARE ALSO OPTIONAL, TO BE ADDED TO TASTE.

LEMON JUICE
2 TABLESPOONS

TAHINI (SESAME PASTE)
3 OR 4 TABLESPOONS

1 GARLIC CLOVE

NOTE:
IF CANNED CHICKPEAS ARE USED, SKIP THE SOAKING AND COOKING STAGE, BUT THEY MUST BE DRIED BEFORE MIXING WITH THE REST OF THE INGREDIENTS.

SALT 1 TEASPOON
CUMIN

OLIVE OIL
1 TABLESPOON
SESAME OIL
APPROX. 1 TABLESPOON

East and Southern Africa

THE JOURNEY OF PULSES TO THE CRADLE OF HUMANITY

In the cultural history of the inhabitants of the savannah and plateau ecosystems, agriculture was a latecomer: hunter-gatherers were the norm in these regions up until some 2 000 years ago. Then the model changed, probably due to the arrival of Bantu tribes from the centre of the continent. The agricultural knowledge and command of foraging that these newcomers brought with them revolutionised the region.

Some palaeobotanical theories place the origin of the *Fabaceae* family squarely in Africa. They even argue that these plants are the most common spontaneously generating species in the continent's jungles and dry forests. There is no doubt that the first human inhabitants of East Africa made use of these pulses as food early on and began to store them, even if their domestication proved impracticable for nomadic tribes. But pulses helped them ensure an adequate diet with low dependence on animal protein. This is evidenced by basic traditional

recipes of high nutritional value that have been passed down from generation to generation, such as the Kenyan Kikuyu's *irio*, a purée of cereals and pulses.

Today, beans of all kinds feature heavily in the regional cuisines of East Africa, both of the *Phaseolus* genus, including black-eyed peas or *adzuki* beans, and other families such as pigeon peas. Not to ignore of course pulses native to the continent like Bambara beans (originally from West Africa) and *lablabs* (*njahi* in the Kikuyu language). In Rwanda, beans are usually cooked in large pots and stored in a variety of ways while in Burundi, pinto beans are a daily staple, often combined with potatoes. Kenyans like to eat them with yucca but, in Uganda, they prefer them with sesame paste, *simsim*, or in stews like their famous *kikomando* recipe.

The region's tribal cuisines combine some unique ingredients with pulses. The Masai, for instance, use cow's milk or blood, and some tribes even use insects such as termites or *mopane* worms, sautéed with lima beans and *sadza* flour, undoubtedly a good source of protein. Another important region is the Horn of Africa, an area that has experienced serious challenges when it comes to food, and where carbohydrates from the omnipresent *ugali* porridge are the perfect complement to dry pulses. The *shahan ful* recipe, a similar dish to the Egyptian *ful medames*, calls for combining them with vitamin-rich raw vegetables and lemon.

South African cuisine has also contributed its fair share of specialities using pulses to African cuisine, such as the hard-to-pronounce Bantu dish, *umngqusho*. The *xhosa* variation was Nelson Mandela's favourite recipe. Influenced by European and Asian cuisine, South Africa boasts a tradition of pulse dishes such as the popular sheep's trotters or pork and beans.

Madagascan cuisine, influenced by the busy trade with America starting in the sixteenth century and with Asia long before then, combines a variety of preparations and products, from pork and *Bambara* bean stews to pots of lima beans with potatoes. Of course, in a country with a Portuguese colonial history like Mozambique, feijoada is a traditional dish. And travelling further east, to the Seychelles archipelago in the Indian Ocean, lentil dishes are commonplace.

The United Republic of Tanzania is distinguished as one of the world's leading producers of dry beans. In fact it's Africa's top producer, well over and above other countries in the region with strong agricultural industries, such as Kenya and Uganda. There, broad beans are an ever-present accompaniment in traditional homemade cooking – along with *ugali*, rice and the vegetable known as *mchicha*, a type of amaranth. They are always prepared simply with just a touch of salt, pepper and, at most, a little tomato. The cooked beans are eaten with minimal garnish and accompanied by animal protein, be it fish or meat.

For the African continent, with its infinite nature reserves made up of savannahs, forests, mountains and beaches, its countless protected animal species and over 120 tribes and scattered ethnic groups, their abundant production of pulses serve not only as an excellent source of food and nutrition for the population, but also as an efficient way of preserving this unique ecosystem.

Tanzania

VERONICA JACKSON: MASAI SENSITIVITY

At the foot of the Ngorongoro Crater in northern Tanzania are nestled the Tanganyika Wilderness Camps. The camps are set within large tracts of land earmarked for long term environmental protection and sustainable development. The aim is to promote green tourism and authentic experiences in an age-old atmosphere. Visitors can enjoy East Africa's rich wildlife and landscape of rare beauty, including areas still inhabited by legendary tribes. Safaris are offered as a bloodless cultural activity. The Ngorongoro Farm House works in partnership with local communities to contribute to their prosperity. Their conviction is that it will only be possible to preserve Africa's wildlife if the local population is involved and benefits from it; hence all of the staff is African.

Veronica Jackson is an example of this dynamic duo of personal and community involvement to promote age-old traditions of Tanzania while creating an exclusive and vibrant hospitality industry. Of Masai origin and the only woman of five siblings, the industrious Veronica has been head chef for the past 15 years at the Kitela Lodge. She oversees the hotel and camp kitchens that are part of the Tanganyika Wilderness Camps initiative.

Born in 1970, Jackson studied in one of the country's handful of cookery schools, the Forozan, in Dar es Salaam. Situated on the Indian Ocean, Dar es Salaam, Tanzania's largest city, was the country's capital until 1996, but during the colonial era it found itself the capital of German East Africa and of Tanganyika (under the British Protectorate), until its independence and the unification of Tanganyika and Zanzibar. Jackson is both expert and enthusiast of the region's traditional cuisine. Notwithstanding her position, she has no qualms about experimenting with new dishes or adapting them to her customer's tastes. After her initial training, she apprenticed

Tanzania

for just one year, before running kitchens in various establishments for 30 more years. Her roles included a stint as head chef at the Masek Tented Camp, a pioneer of Africa's eco-friendly camp tourism, situated in the Ngorongoro Conservation Area, in southern Seregenti.

In addition to her role as executive chef, Veronica is involved in the management and public relations of the resort where, as the chief in all culinary matters, she is indispensable. At the Tanganyika Wilderness Camps, daily meals feature gourmet experiences with ethnic recipes adapted to both meet conventional tastes of foreign visitors and incorporate contemporary western and eastern culinary trends.

The produce is ultra-local and used in daily meals at the camps' restaurants and excursions. Here, the tasty, ancestral tradition of pulses features as part of a local, healthy and balanced diet. "Pulses are the staple food of the people of Tanzania," Veronica explains, "and demand is growing in the country, so we must make sure that they are produced. We have to expand and facilitate growing to guarantee supply and distribution so that the wealth of pulse varieties is available to everyone."

In fact, in addition to the supplies from markets in Arusha, Karatu and Mto wa Mbu (well worth a visit to discover the region's unique products), the Tanganyika Wilderness Camps' own *shambas* (kitchen gardens) produce conventional and native pulses that illustrate the diversity of the ethnic cuisine of the Masai people and other remote tribes that survive to this day, like the hunter-gatherers of the Hadzabe, Datoga and Barabaig.

Veronica Jackson is responsible for the growing cycles of the gardens that surround the camps, where the nearby communities are permitted to plant pulses in the coffee plantations for their own use. The many local varieties of pulses grown there enrich the culinary experience, and their delicious and exotic names whet the appetite: *kunde kunde, mbaazi, maharagwe soya, maharagwe mabichi, dengu, choroko, nyayo maharagwe, ngwara mea…*

Chef Veronica Jackson selecting pulses in the local market of Karatu (Tanzania). The remaining images show the chef in her garden patch and in the kitchen of the restaurant in *Kitela Lodge*.

Maharagwe

TRADITIONAL PLATE

SERVES 4

HOW TO PREPARE:

1) Boil kidney beans in a pot of water until soft. Drain keeping the water aside.

2) Place onion and garlic in another pot and fry in oil until tender.

3) Add the minced meat and cook for 10 minutes.

SOAKED APPROX. 10 HOURS

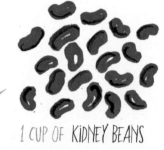

1 CUP OF KIDNEY BEANS

CHOPPED CORIANDER

4) ADD THE TOMATO, CARROT AND CORIANDER AND LIGHTLY FRY FOR 5 MINUTES.

5) ADD THE COOKING WATER FROM THE BEANS AND BOIL FOR 2 MINUTES.

6) REMOVE FROM HEAT, MIX WITH THE KIDNEY BEANS AND SERVE.

=INGREDIENTS:=

1/2 KG OF MINCED **MEAT**

2 TOMATOES CHOPPED

1 ONION CHOPPED

2 CARROTS CHOPPED

2 GARLIC CLOVES CRUSHED

OIL FOR FRYING

MAHARAGWE YA NAZI
(BEANS IN COCONUT MILK)

SOAKED APPROX.

10 HOURS

2 CUPS OF DRIED **KIDNEY BEANS**

OIL FOR FRYING

1/2 TABLESPOON OF GROUND **CINNAMON**

2 **ONIONS** CHOPPED

SOME **CARDAMOM SEEDS**

1 TABLESPOON OF **CURRY** POWDER

2 GARLIC CLOVES CRUSHED AND CHOPPED

1 TEASPOON OF **SALT**

2 TOMATOES CHOPPED

1 CUP OF **COCONUT MILK**

2 TABLESPOONS OF **CASTOR SUGAR**

NOTE:

MAHARAGWE YA NAZI (BEANS IN COCONUT MILK) IS USUALLY SERVED WITH CHAPPATI (TOASTED, UNLEAVENED INDIAN BREAD) AND RICE.

How to prepare: Serves 4

1 FRY ONIONS IN A PAN FOR 5 MINUTES.

2 ADD THE TOMATOES AND SAUTÈ UNTIL TENDER.

3 ADD THE BEANS, COVER WITH WATER AND COOK UNTIL SOFT.

4 ADD THE GARLIC, CARDAMOM SEEDS, CURRY POWDER, CINNAMON AND SALT AND CONTINUE COOKING, STIRRING THE MIXTURE FOR 1 MINUTE.

5 STIR IN COCONUT MILK FOLLOWED BY SUGAR, STIR, COVER AND COOK UNTIL EVERYTHING IS TENDER.

6 WHEN THE COCONUT MILK IS THE SAME CONSISTENCY AS THE BEANS, IT IS READY FOR PARSLEY GARNISH AND SERVING.

PARSLEY

Makande

SWEETCORN AND BEAN STEW

SERVES 4

VEGETARIAN PLATE

1) DRAIN THE KIDNEY BEANS, COVER THEM IN WATER. ADD THE TOMATOES AND CARROTS AND BOIL ON A HIGH HEAT FOR 20 MINUTES.

LOWER THE HEAT AND CONTINUE TO COOK FOR AROUND 1 HOUR, UNTIL TENDER. ADD MORE WATER DURING COOKING IF NECESSARY.

2) PLACE THE BEANS IN A CLEAN POT WITH THE SWEETCORN, ONION, GARLIC, COCONUT CREAM, SALT AND PEPPER.

3) ADD BOILING STOCK OR WATER AND SIMMER FOR 20 MINUTES TO DISSOLVE THE COCONUT CREAM.

4) ADJUST SEASONING AND SERVE HOT.

75 G OF COCONUT CREAM

2 CUPS OF SWEET CORN

2 GARLIC CLOVES, CRUSHED

2 ONIONS, CHOPPED

2 TOMATOES, CHOPPED

2 CARROTS, MASHED

SALT AND PEPPER TO TASTE!!

NOTE:

THIS IS A TRADITIONAL TANZANIAN DISH THAT CAN BE SERVED WITH MEAT, FISH OR SIMPLY WITH SALAD.